T0256000

SpringerBriefs in Applied Sciences and Technology

SpringerBriefs present concise summaries of cutting-edge research and practical applications across a wide spectrum of fields. Featuring compact volumes of 50–125 pages, the series covers a range of content from professional to academic.

Typical publications can be:

- A timely report of state-of-the art methods
- An introduction to or a manual for the application of mathematical or computer techniques
- A bridge between new research results, as published in journal articles
- A snapshot of a hot or emerging topic
- An in-depth case study
- A presentation of core concepts that students must understand in order to make independent contributions

SpringerBriefs are characterized by fast, global electronic dissemination, standard publishing contracts, standardized manuscript preparation and formatting guidelines, and expedited production schedules.

On the one hand, **SpringerBriefs in Applied Sciences and Technology** are devoted to the publication of fundamentals and applications within the different classical engineering disciplines as well as in interdisciplinary fields that recently emerged between these areas. On the other hand, as the boundary separating fundamental research and applied technology is more and more dissolving, this series is particularly open to trans-disciplinary topics between fundamental science and engineering.

Indexed by EI-Compendex, SCOPUS and Springerlink.

More information about this series at http://www.springer.com/series/8884

Aydin Azizi

Applications of Artificial Intelligence Techniques in Industry 4.0

 Springer

Aydin Azizi
Department of Engineering
German University of Technology in Oman
Muscat, Oman

ISSN 2191-530X   ISSN 2191-5318 (electronic)
SpringerBriefs in Applied Sciences and Technology
ISBN 978-981-13-2639-4   ISBN 978-981-13-2640-0 (eBook)
https://doi.org/10.1007/978-981-13-2640-0

Library of Congress Control Number: 2018955449

This Springer imprint is published by the registered company Springer Nature Singapore Pte Ltd.
The registered company address is: 152 Beach Road, #21-01/04 Gateway East, Singapore 189721, Singapore

To My Family

Contents

1 **Introduction** 1
  1.1 Overview 1
  1.2 Aims and Objectives 2
  1.3 Methodology 3
  1.4 Structure of Book 4
  References 5

2 **Modern Manufacturing** 7
  2.1 Internet of Thing 7
  2.2 Radio Frequency Identification Technology 10
      2.2.1 Introduction 10
      2.2.2 Components of RFID System 12
  References 16

3 **RFID Network Planning** 19
  3.1 Overview 19
  3.2 Mathematical Modeling 20
      3.2.1 Coverage 21
      3.2.2 Redundant Antennas 23
      3.2.3 Interference 23
      3.2.4 Transmitted Power 24
  References 24

4 **Hybrid Artificial Intelligence Optimization Technique** 27
  4.1 Overview 27
  4.2 Methodology 29
      4.2.1 Redundant Antenna Elimination Algorithm 30
      4.2.2 Ring Probabilistic Logic Neural Networks 34
  4.3 Genetic Algorithm 44
  References 46

5 Implementation   49
  5.1   Overview   49
  5.2   Working Area   49
    5.2.1   Static Working Area   49
    5.2.2   Dynamic Working Area   50
  5.3   Parameters of the Proposed Hybrid Algorithm   52
    5.3.1   Population of the Possible Answers   52
    5.3.2   Fitness Function   53
  5.4   Results   54
    5.4.1   Static Working Area   54
    5.4.2   Dynamic Working Area   57
  5.5   Conclusion   60
  Reference   61

List of Figures

Fig. 1.1   Proposed methodology of optimizing RNP problem 3

Fig. 1.2   Structure of this book 5

Fig. 2.1   Definition of Internet of Things [4] 8

Fig. 2.2   Internet of Things: intelligent systems framework [7] 8

Fig. 2.3   IoT model for manufacturing and industrial automation [7] 9

Fig. 2.4   Connected enterprise [7] 10

Fig. 2.5   Integrated equipment and appliances [7] 11

Fig. 2.6   The digital retail store [11] 11

Fig. 2.7   Components of an RFID system [18] 13

Fig. 2.8   Interactions between components of RFID system [19] 13

Fig. 2.9   RFID tag [21] 13

Fig. 2.10   RFID tag with printed barcode on it [22] 14

Fig. 2.11   RFID active tag [24] 15

Fig. 2.12   RFID passive tag [25] 15

Fig. 3.1   RFID technology components [2] 20

Fig. 3.2   Steps of defining an RNP 21

Fig. 3.3   Working area and distributed 20 tags 21

Fig. 3.4   Example of tag coverage, if $PT_{a_1,t_n} \geq T_{t_n} \wedge PA_{t_n,a_3} \geq T_{a_3}$,
then $C_v (1) = 1$ [3] 22

Fig. 4.1   Proposed hybrid optimization technique 28

Fig. 4.2   RNP optimization process 29

Fig. 4.3   Proposed hybrid artificial intelligence optimization
process [16] 31

Fig. 4.4   Flowchart of the proposed redundant antenna elimination
algorithm [16] 33

Fig. 4.5   Biological neuron [18] 35

Fig. 4.6   Example of artificial neural networks [18] 35

Fig. 4.7   A RAM neuron [28] 36

Fig. 4.8   A RAM discriminator [30] 37

Fig. 4.9   Flowchart of training algorithm of RAM neural network 38

Fig. 4.10   Probabilistic Logic Neuron (PLN) [16].   39
Fig. 4.11   Pyramidal PLN neural network structure [16].   39
Fig. 4.12   Example of a PLN neural network structure [16]   40
Fig. 4.13   RPLNN structure [16].   41
Fig. 4.14   Flowchart of the proposed RPLNN algorithm to optimize
            RNP   42
Fig. 4.15   Example of roulette wheel [37].   45
Fig. 4.16   **a** Two-point crossover, **b** single-point crossover [16].   45
Fig. 4.17   Example of mutation [37].   45
Fig. 5.1   Proposed static working area   50
Fig. 5.2   FMS [1]   50
Fig. 5.3   Proposed dynamic working area.   51
Fig. 5.4   Proposed population of possible answers   53
Fig. 5.5   Example of encoded answer in population of answers   53
Fig. 5.6   Calculated number of deployed antennas in the network
            by the proposed hybrid algorithm and GA in each iteration   55
Fig. 5.7   Calculated coverage of the RFID network by the proposed
            hybrid algorithm and GA in each iteration   55
Fig. 5.8   Calculated ITF of the network by the proposed hybrid
            algorithm and GA in each iteration.   56
Fig. 5.9   Normalized fitness function values of the proposed hybrid
            algorithm and GA in each iteration.   56
Fig. 5.10   Optimized RFID network   57
Fig. 5.11   Calculated number of deployed antennas in the network
            by the proposed hybrid algorithm and GA in each iteration   58
Fig. 5.12   Calculated coverage of the RFID network by the proposed
            hybrid algorithm and GA in each iteration   58
Fig. 5.13   Calculated ITF of the network by the proposed hybrid
            algorithm and GA in each iteration.   59
Fig. 5.14   Normalized fitness function values of the proposed hybrid
            algorithm and GA in each iteration.   59
Fig. 5.15   Optimized RFID network   60

Abstract

Recent advances in technology and modern manufacturing industry as an important part of Industry 4.0 have created a great need to model the behavior of manufacturing systems. Nowadays, this need with the developments in computer technology and software engineering can be addressed by modern computational techniques. Artificial intelligence (AI) is one of the well-known advanced computational techniques which is growing fast and has been utilized to model, control, and optimize different disciplines of engineering, in which manufacturing industry is no exception.

Obtaining real-time information has a great value in different fields of manufacturing industry such as flexible manufacturing systems, inventory management, and supply chain management. One of the developing technologies which has been utilized to identify and track parts and objects in the manufacturing industry is radio-frequency identification (RFID) system. An RFID system contains three major components: tags which are mounted at the parts needed to be tracked, antenna to read tags, and computer as a middleware.

Several challenges have arisen due to adopting RFID in the manufacturing industry environment. One of these challenges which has been a research area of many scientists is known as RFID network planning (RNP) problem. Mainly, RNP deals with calculating the number of antennas which should be deployed in the RFID network to achieve full coverage of the tags which are needed to be read. Several different optimization techniques have been used to optimize RNP, but many of them are complex and inefficient. The goal of this book is to present and evaluate a way of modeling and optimizing nonlinear RNP problem utilizing artificial intelligence techniques. The research has been developed by utilizing artificial neural network (ANN) models, computational artificial intelligence algorithms and mathematical models of RFID networks planning (RNP) to develop an efficient artificial intelligence paradigm to model and optimize RFID networks.

This effort has led to proposing a novel artificial intelligence algorithm which has been named hybrid artificial intelligence optimization technique to perform the optimization of RNP as a hard learning problem. This hybrid optimization technique has been made of two different optimization phases. The first phase is optimizing RNP by redundant antenna elimination (RAE) algorithm, and the second phase which completes the RNP optimization process is ring probabilistic logic neural networks (RPLNNs).

The proposed hybrid paradigm has been explored using a flexible manufacturing system (FMS), and the results are compared with the well-known evolutionary optimization technique, namely genetic algorithm (GA), to demonstrate the feasibility of the proposed architecture successfully.

Keywords Industry 4.0 · Manufacturing Industry · Flexible Manufacturing System (FMS) · Radio Frequency Identification (RFID)
RFID Network Planning (RNP) · Artificial Intelligence · Artificial Neural Networks (ANNs) · Hybrid Artificial Intelligence Algorithm
Redundant Antenna Elimination (RAE) · Probabilistic Logic Neural Networks (RPLNNs) · Genetic Algorithm (GA)

Chapter 1
Introduction

1.1 Overview

The steady-state industry status has been changed to dynamic industry by the industrial revolution, so manufacturers have been pushed by the global market to reconsider their conventional manufacturing methods. Modern manufacturing needs new manufacturing operations, and effective factory management has a great value in this area. Recent advances in technology and modern industrial engineering systems from production to transportation have created a great need to track and identify the materials, products, and even live subjects [1].

Radio Frequency Identification (RFID) technology is a reliable and efficient solution to this tracking and identifying the problem. RFID technology is known as an automatic identification technology as it uses wireless radio frequency waves which are produced by an electromagnetic field to transfer data to track and identify objects. This technology can be implemented in different fields such as tracking and identifying patients in hospitals [2], warehouse items tracking [3], tracking pallets and cases in shipment [3], monitoring production line [4], and supply chain management [5].

In many applications, the deployment of RFID systems has generated an RFID network planning (RNP) problem [9] which needs to be resolved, in order to efficiently operate a large-scale network. However, RNP is one of the most challenging problems that must meet many requirements of the RFID system [6–8]. In general, the RNP aims to optimize a set of objectives (coverage, load balance, economic efficiency and interference between antennas, etc.) simultaneously; this is achieved by adjusting the control variables (the coordinates of the antennas, the number of antennas, etc.) of the system. As a result, in a large-scale deployment environment, the RNP problem is a high-dimensional nonlinear optimization problem that has a vast number of variables and uncertain parameters.

Tracking and identifying objects in these applications require the deployment of several RFID antennas in the RNP, and the numbers of these antennas are calculated

© The Author(s), under exclusive licence to Springer Nature Singapore Pte Ltd. 2019
A. Azizi, *Applications of Artificial Intelligence Techniques in Industry 4.0*,
SpringerBriefs in Applied Sciences and Technology,
https://doi.org/10.1007/978-981-13-2640-0_1

using a mathematical model [9–11]. In the past, one of the typical ways to address the RNP problem was using a trial and error approach, which was an inaccurate and inefficient solution for such an important issue. In addition, this approach could only be used in small-scale RFID network planning problems [12]. Nowadays with the developments in computer technology and software engineering, the conventional trial and error approach has been replaced with modern computational techniques that provide such important criteria like coverage of objects, collision of antennas, and number of antennas. Computational evolutionary techniques such as Artificial Neural Networks [13, 14], Fuzzy Logic [15], Genetic Algorithms (GA) [14, 16], Particle Swarm Optimization (PSO) [14, 17], Differential Evolution (DE) [18], and hierarchical artificial bee colony algorithm [19] are points of interest for many scientists working with the RNP problem.

The aim of this book is to present and evaluate a novel way of optimizing nonlinear RNP problems utilizing artificial intelligence techniques. The research has been developed by utilizing Artificial Neural Network models (ANN), computational artificial intelligence algorithms and mathematical modeles of RFID network planning (RNP) to develop an efficient artificial intelligence paradigm to optimize nonlinear RNP problems. Starting from introducing existing ANN models, it defines which structure is required in order to optimize functions. Different artificial intelligence algorithms, which can satisfy the required capabilities for optimizing of defined RFID network planning problem that can be represented as mathematical models, are presented and discussed. This effort has led to proposing a novel artificial intelligence algorithm which has been named hybrid artificial intelligence optimization technique to perform the optimization of RNP as a hard learning problem. The proposed hybrid optimization technique has been made of two different optimization phases. The first phase is optimizing RNP by Redundant Antenna Elimination (RAE) algorithm and the second phase which completes the RNP optimization process is Ring Probabilistic Logic Neural Networks (RPLNN).

1.2 Aims and Objectives

The objective of this book is to present and evaluate a novel way of optimizing RNP problem as an important part of manufacturing industry utilizing artificial intelligence techniques. The research has been developed utilizing Artificial Neural Network models (ANN) to bind together the computational artificial intelligence algorithms and mathematical models of RFID network planning (RNP) to develop an efficient artificial intelligence paradigm to optimize deployed number of antennas in RFID network based on the criteria of defined RNP as a nonlinear engineering problem.

Starting from defining radio frequency identification systems, it defines the challenges of establishing an efficient RFID network and introduces existing RNP models; it is followed by introducing the existing artificial neural networks models; it defines which structure is required to optimize nonlinear RNP function. Different artificial intelligence algorithms, which can satisfy the required capabilities for optimizing of defined RNP problems which can be represented as mathematical models, are presented and discussed. This effort has led to the utilizing of a novel artificial intelligence algorithm which is named hybrid artificial intelligence optimization technique to perform the optimization of RNP as a hard learning problem. The proposed hybrid optimization technique has been made of two different optimization phases. The first phase is optimizing RNP by Redundant Antenna Elimination (RAE) algorithm and the second phase which completes the RNP optimization process is Ring Probabilistic Logic Neural Networks (RPLNN).

The ultimate goal of this research is to introduce ring probabilistic logic neuron as a time efficient and reliable optimization technique to solve RFID network planning problem and design a cost-effective RFID network by minimizing the number of embedded RFID antennas in the network, minimizing collision of antennas and maximizing coverage area of objects.

The proposed hybrid artificial intelligence paradigm has been explored using a flexible manufacturing system (FMS) and the results are compared with Genetic Algorithm (GA) optimization technique to demonstrate the feasibility of the proposed architecture successfully.

1.3 Methodology

The proposed methodology of this book contains three phases: problem identification, research and development, and implementation (see Fig. 1.1).

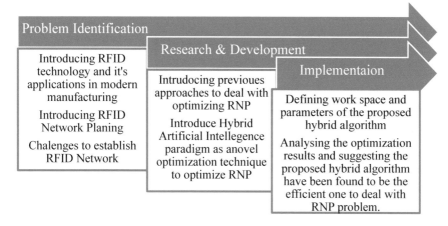

Fig. 1.1 Proposed methodology of optimizing RNP problem

The main contributions of this book in detail are as follows:

1. Problem Identification

    (i)   Introducing RFID systems as application which has been used in manufacturing industry.
    (ii) Sufficiently reviews the previous work on RFID network planning.
    (iii) Identifying the existing challenges to establish an efficient RFID network.

2. Research and Development

    (i)   Sufficiently reviews the previous work on for optimizing RNP.
          (1) Introduces artificial neural network as computational intelligent technique in detail to deal with RNP, including certain extensions and applications.
          (2) Investigate and introduce different artificial neural networks models.
    (ii) Proposes hybrid artificial intelligence as a novel technique to deal with optimizing RNP.
          (1) Introduces Redundant Antenna Elimination (RAE) algorithm as the first part of the proposed hybrid algorithm.
          (2) Introduces Ring Probabilistic Logic Neural Networks (RPLNN) as the second part of the proposed hybrid algorithm.

3. Implementation

    (i)   Defines the working area of RNP.
    (ii) Implements proposed technique in flexible manufacturing system (FMS).
    (iii) Comparative analysis performance of the proposed algorithm with genetic algorithm optimization technique.
    (iv) Explains some current difficulties and problems based on the summary and conclusion of the book and proposes future solutions (Fig. 1.2).

1.4 Structure of Book

The rest of this research is organized as follows: Chap. 2 is an introduction in modern manufacturing, Internet of Things technology and its applications beside introducing radio frequency identification systems as one of the important adopted technologies by Internet of Things. Chapter 3 is an introduction to RFID network planning and introduces recent mathematical models of RFID network planning. A novel hybrid optimization technique to optimize the RNP has been introduced in Chap. 4, and results of implementing the proposed approach on a flexible manufacturing system have been discussed in Chap. 5.

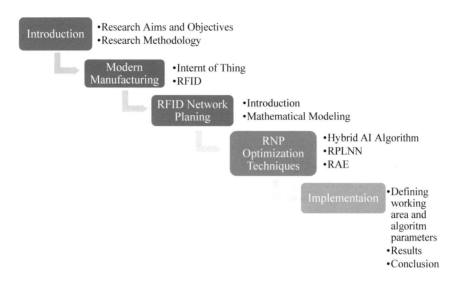

Fig. 1.2 Structure of this book

References

1. S.S. Kamble, A. Gunasekaran, S.A. Gawankar, Sustainable Industry 4.0 framework: a systematic literature review identifying the current trends and future perspectives. Process Saf. Environ. Prot. **117**, 408–425 (2018)
2. R.R. Thapa et al., in *Application of RFID Technology to Reduce Overcrowding in Hospital Emergency Departments*, ed. by A.G. Nilsson, R. Gustas, G. Wojtkowski, W. Wojtkowski, S. Wrycza, J. Zupancic, Advances in Information Systems Development (Springer, 2018), pp. 17–32
3. M. Ma, P. Wang, C.-H. Chu, Redundant reader elimination in large-scale distributed RFID networks. IEEE Internet Things J (2018)
4. M.S. Altaf et al., Integrated production planning and control system for a panelized home prefabrication facility using simulation and rfid. Autom Constr **85**, 369–383 (2018)
5. S. Luthra et al., in *Internet of Things (IoT) in Agriculture Supply Chain Management: A Developing Country Perspective*, ed. by Y.K. Dwivedi, N.P. Rana, E.L. Slade, M.A. Shareef, A.C. Simintiras, B. Lal, Emerging Markets from a Multidisciplinary Perspective (Springer, 2018), pp. 209–220
6. W. Shi et al., Optimizing Directional Reader Antennas Deployment in UHF RFID localization system by using a MPCSO algorithm. IEEE Sens. J. **18**(12), 5035–5048 (2018)
7. A. Raghib, B.A. El Majd, B. Aghezzaf, in *An Optimal Deployment of Readers for RFID Network Planning Using NSGA-II*, ed. by A. Lionel, E.L. Talbi, F. Yalaoui, Recent Developments in Metaheuristics (Springer, 2018), pp. 463–476
8. M. Munkailu, S. Sani, A. Tekanyi, Development of a sparse RFID reader deployment algorithm for effective RFID network planning. Niger. J. Technol. **37**(3), 779–785 (2018)
9. P. Vestenický, M. Vestenický, in *2018 ELEKTRO*, ed. by P. Vestenický, M. Vestenický, Mathematical Model of Non-linear RFID Reader—Transponder System (IEEE, 2018)
10. F. Amato, H.M. Torun, G. Durgin, RFID backscattering in long-range scenarios. IEEE Trans. Wireless Commun. (2018)
11. B.W. Podaima et al., Modeling RFID tracking in healthcare. CMBES Proc. **33**(1) (2018)

12. A. Suriya, J.D. Porter, in *Proceedings of IIE Annual Conference*, ed. by A. Suriya, J.D. Porter, RFID Network Modeling and Optimization for Inventory Management (Institute of Industrial and Systems Engineers (IISE), 2012)
13. R.V. Aroca et al., Calibration of passive UHF RFID tags using neural networks to measure soil moisture. J. Sens. (2018)
14. A. Azizi, A. Vatankhah Barenji, M. Hashmipour, Optimizing radio frequency identification network planning through ring probabilistic logic neurons. Adv. Mech. Eng. **8**(8), 1687814016663476 (2016)
15. C.O. Chan, H. Lau, Y. Fan, in *2018 International Conference on Artificial Intelligence and Big Data (ICAIBD)*, ed. by C.O. Chan, H. Lau, Y. Fan, *IoT Data Acquisition in Fashion Retail Application: Fuzzy Logic Approach* (IEEE, 2018)
16. W. Zhu, M. Li, RFID reader planning for the surveillance of predictable mobile objects. Procedia Comput. Sci. **129**, 475–481 (2018)
17. X. Cai, L. Ye, Q. Zhang, Ensemble learning particle swarm optimization for real-time UWB indoor localization. EURASIP J. Wireless Commun. Netw. **2018**(1), 125 (2018)
18. S.K. Goudos, in *Encyclopedia of Information Science and Technology*, 4th edn, ed. by S.K. Goudos, *Optimization of Antenna Arrays and Microwave Filters Using Differential Evolution Algorithms* (IGI Global, 2018), pp. 6595–6608
19. L. Ma et al., Two-level master-slave RFID networks planning via hybrid multiobjective artificial bee colony optimizer. IEEE Trans. Syst. Man Cybern. Syst. (2017)

Chapter 2
Modern Manufacturing

2.1 Internet of Thing

The steady-state industry status has been changed by the industrial revolution to dynamic industry, so manufacturers have been pushed by the global market to reconsider their conventional manufacturing methods. Modern manufacturing needs new manufacturing operations, and effective factory management has a great value in this area [1]. Smart manufacturing is a powerful concept which can be addressed as an answer to needs of modern manufacturing by utilizing and adding high-tech products such as sensors, software, and wireless connectivity to required products [2]. Overall, utilizing high-tech equipment in manufacturing to optimize the manufacturing methods results in exploring a new concept, which is known as Internet of Things (IoT) [3] (Fig. 2.1).

The Internet of Things (IoT) can be defined as interaction between technologies which includes smart objects, machine to machine communication, radio frequency technologies, and a central hub of information to monitor the status of physical objects, capturing meaningful data, and communicating that information through IP networks to software applications. In IoT to make objects detectable to monitor and collect required data from them, they are equipped with an Auto-ID technology. Utilizing this technology enables the users to analyze the collected data which can contain the information such as temperature, changes in quantity, or other types of information through wireless communication and make efficient and accurate decisions [5].

In recent years with advances in technology IoT has started to adopt and utilize a new technology which is named Radio Frequency Identification (RFID), continuously increases its market share, replacing traditional barcode technology and allows for the development of new applications [6] (Fig. 2.2).

RFID technology can be defined as a powerful innovative gadget which has been adopted in development of IoT. RFID technology as an application in IoT with accepted standards across industries widely has been adopted in different

© The Author(s), under exclusive licence to Springer Nature Singapore Pte Ltd. 2019   7
A. Azizi, *Applications of Artificial Intelligence Techniques in Industry 4.0*,
SpringerBriefs in Applied Sciences and Technology,
https://doi.org/10.1007/978-981-13-2640-0_2

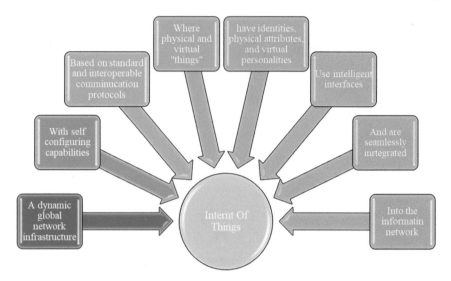

Fig. 2.1 Definition of Internet of Things [4]

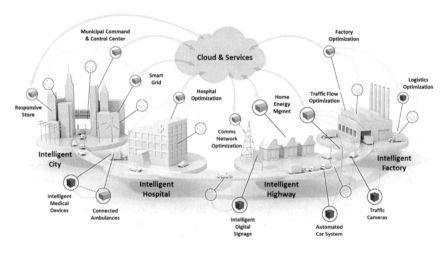

Fig. 2.2 Internet of Things: intelligent systems framework [7]

manufacturing industry section; we are represented some of the most important examples as below:

- Supply Chain Management

Employing RFID systems as an IoT application has a great importance in supply chain management as a valuable part of modern manufacturing and industrial automation. With the aid of using RFID system, managers will be able to track, monitor, and control their products in real time from the status of being as row material to rotation

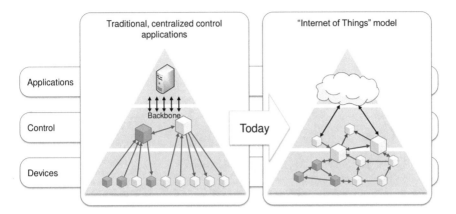

Fig. 2.3 IoT model for manufacturing and industrial automation [7]

as final products in shelves and warehouses. They will have all required data related to shipments, location, temperature, pressure, and even days until expiration [8] (Fig. 2.3).

- Smart Manufacturing

A smart and efficient manufacturing can be achieved by utilizing IoT which will connect the factory to the applications run around the production. On the other hand, it can be said that in smart manufacturing, the manufacturer is enabled to include suppliers, production logistics, and even maintenance [9]. This means that the services of IoT-enabled manufacturing system will be in a shared physical world rather than restricted in a physical system (Fig. 2.4).

- Home Automation Industry

Nowadays, home automation industry by adopting RFID in IoT is closer to its goal which is interconnected home application. Utilizing RFID technology to enable home residents using a remote device to control all home electronic devices and appliances has been the point of research of many organizations. This concept has been introduced by smart products such as smart air conditioners, smart thermostats, etc., which can be monitored and controlled from distance just by an application of a smartphone [10] (Fig. 2.5).

- Intelligent Shopping

People in terms of consumers can manage their time daily life in a more efficient way, and also they can save money by utilizing RFID technology as an application of the Internet of Things application. A real-time grocery list can be generated automatically for consumers and list the items that will be consumed or expired in near future date. So, it prevents people to buy the items which are not necessary and it will result in reducing the waste [11]. Also, utilizing RFID as an IoT application by

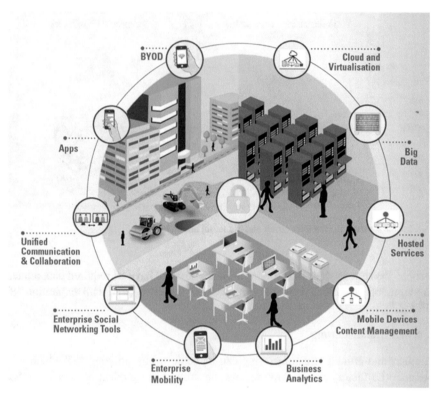

Fig. 2.4 Connected enterprise [7]

retailers will enable them to collect real-time data from their inventory, resources, products, etc., to have a better overview of their chain of customers, employees (Fig. 2.6).

After giving a brief review about the Internet of Things as an important part of manufacturing industry and introducing radio frequency identification technology as a powerful gadget and application in IoT in the next sections of this chapter, RFID technology and its components have been introduced in detail, and the establishment of an RFID network will be discussed.

2.2   Radio Frequency Identification Technology

2.2.1   Introduction

Obtaining real-time information has a great value in different fields of manufacturing industry such as flexible manufacturing systems, inventory management, and supply chain management. Recent advances in technology and modern industrial engineer-

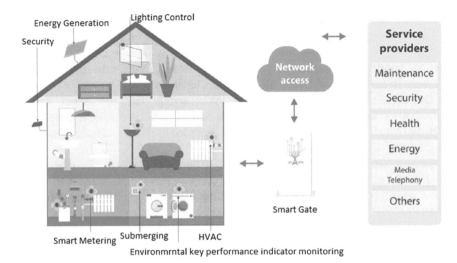

Fig. 2.5 Integrated equipment and appliances [7]

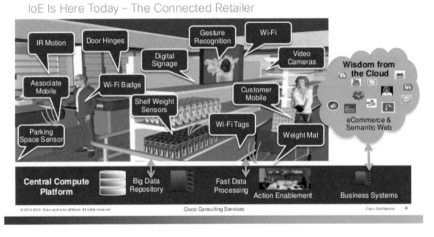

Fig. 2.6 The digital retail store [11]

ing systems from production to transportation have created a great need to track and identify the materials, products, and even live subjects [12]. One of the developing technologies which have been utilized as an application in Internet of Thing (IoT) to identify and track parts and objects in manufacturing industry is Radio Frequency Identification (RFID).

Radio Frequency Identification (RFID) technology is an Auto-ID technology which is adopted by IoT as a reliable and efficient solution to the problem of object tracking and identifying. RFID technology is known as an automatic identification technology as it uses wireless radio frequency waves which are produced by an electromagnetic field to transfer data to track and identify objects. Utilizing this technology enable users to analyze the collected data which can contain information of the objects such as temperature, changes in quantity, or other types of information through wireless communication and make efficient and accurate decisions [13].

It can be said that modern manufacturing has been changed by impact of utilizing radio frequency identification. These changes have wide range and include but not limited to the manner of the tracking objects not even in small and medium enterprises also the boundaries of tracking of the objects upgraded to global scale and interaction of products with production environment [14].

In past, barcode technology had been used to track and identify objects and items in different manufacturing fields; however, utilizing these applications was cheap but it should be mentioned that since collecting data had been done manually by a barcode reader, it was not an easy and efficient task. The other important drawback of using barcode is the nature of storing the information of the object just in the time of reading, so barcodes do not provide manufacturer required data in real time, and they are capable of recording data just in time, so it makes the barcode technology a non effective technology in modern manufacturing [15].

Nowadays, this technology has been replaced with RFID technology and has been used as an effective application in industries and it has been in point of research interest of many researchers. Windmann et al. [16] investigated RFID importance in production and operation management. Irani et al. [17] proposed the framework for presenting research obstacles related to RFID technology. A comprehensive research about transferring real-time information using RFID technology in value adding chain component has been given by Bibi et al. [12]. In general, as it is mentioned before, this technology can be implemented in different fields such as tracking and identifying patients in hospitals, warehouse items tracking, tracking pallets and cases in shipment, monitoring production line, and supply chain management.

2.2.2 Components of RFID System

An RFID system is made up of two major parts: hardware part and software part. Hardware part consists of tags, readers, antennas, and software part that includes middleware which can be defined also as computer unit [18]. It should be noted that the antenna can be a part of the reader, meaning more than one antenna can be connected to one reader (see Fig. 2.7).

RFID is the advanced technology in data process. Required data from objects are sent by tags to readers; antennas of readers receive these data and send to a host computer as middleware to process for further implementations (see Fig. 2.8).

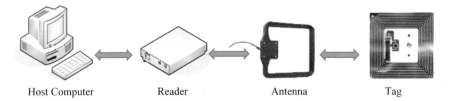

Fig. 2.7 Components of an RFID system [18]

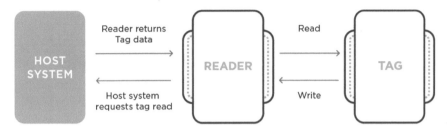

Fig. 2.8 Interactions between components of RFID system [19]

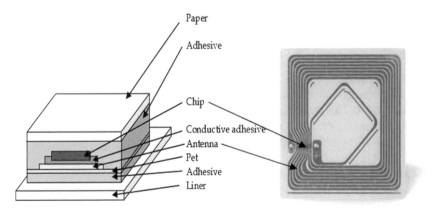

Fig. 2.9 RFID tag [21]

2.2.2.1 RFID Tags

One of the major parts of an RFID system is tag. All the object data are sent by tags to readers; it should be mentioned that tags are made up from microchips which are attached to antennas, and data are sent by these antennas in the form of electromagnetic waves which are known as RF signals (see Fig. 2.9) [20].

RFID tags have different shapes, we can see some of them in daily life in stores and markets as smart labels which include printed barcode on them (see Fig. 2.10), in the form of simple tags which are mounted inside a carton, smart cards.

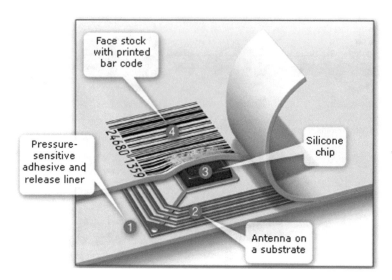

Fig. 2.10 RFID tag with printed barcode on it [22]

In general, RFID tags are divided into three major groups which are passive tags, semi-passive, and active tags [4].

(a) Active Tags

Active tags have their own power source, so they can send and broadcast their own signals. This source of energy can be a battery, PV cells or other sources. From the point of view of the range of broadcasting, active tags have longer range than passive tags because they have their own source of energy to broadcast [23] (Fig. 2.11).

(b) Passive Tags and Semi-passive Tags

Passive tags do not have their own power source, so they cannot send and broadcast their own signals. A passive tag receives signal which has been sent by a reader; this signal contains energy which can be absorbed by microchip circuit of passive tag, so from then passive tag enables to reflect the absorbed signal to reader. It should be mentioned that there is an energy loss due to this reflection and the resultant will be a lower read range of the passive tag [4] (Fig. 2.12).

2.2.2.2   RFID Readers

RFID readers are another major parts of RFID system which has the role of communication with RFID tags. RFID readers make this communication by sending and receiving RF signals back from RFID tags through their antennas. The antenna of a RFID reader is divided into two types: internal antennas and external antennas [18].

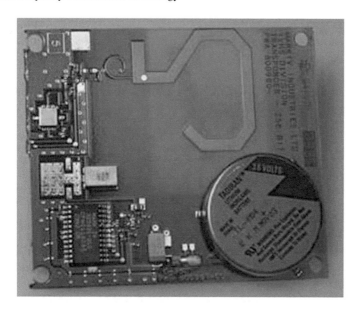

Fig. 2.11 RFID active tag [24]

Fig. 2.12 RFID passive tag [25]

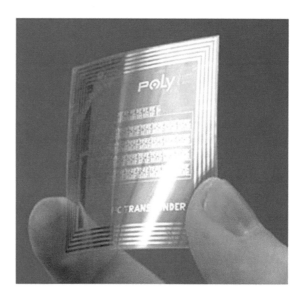

RFID readers which have external antennas can be connected to several antennas; each connection is applicable through a port. It should be mentioned that readers with external antennas can be deployed with up to eight antenna ports. Readers can have different ports such as USB ports, Wi-Fi ports, serial ports, and input/output ports that enable connection to external devices [26].

2.2.2.3   RFID Antennas

The antenna can be a part of the tag or reader and can be defined as a conductive element which tags, and readers can transmit and receive data through that. The antennas can have different shapes but all of them serve as one purpose which is transmitting of radio waves [27].

2.2.2.4   RFID Middleware

RFID middleware can be defined as software part of the RFID system which can be inhabited on a server to filter collected data of RFID readers and may to pass on useful collected data to enterprise applications [28].

Some of the RFID middlewares not only have the function of filtering data, also they can manage and monitor RFID readers. These kind of middlewares configure the RFID readers and monitor their functionality, additionally if it is needed the middleware sends necessary updates to update readers [29].

After giving an introduction to modern manufacturing and RFID technology as an important application which has been adopted in IoT as part of modern manufacturing, the next chapter will introduce mathematical modeling of establishing an RFID network.

References

1. A. Azizi, P. Yazdi, A. Humairi, Design and fabrication of intelligent material handling system in modern manufacturing with industry 4.0 approaches. Int. Rob. Auto. J. **4**(3), 186–195 (2018)
2. A. Azizi, P. Yazdi, A. Humairi, Applications of control engineering in industry 4.0: utilizing internet of things to design an agent based control architecture for smart material handling system. Int. Rob. Auto. J. **4**(4), 253–257 (2018)
3. A. Al-Fuqaha et al., Internet of things: A survey on enabling technologies, protocols, and applications. IEEE Commun. Surv. Tutorials **17**(4), 2347–2376 (2015)
4. A. Azizi, *Introducing a novel hybrid artificial intelligence algorithm to optimize network of industrial applications in modern manufacturing*. Complexity **2017** (2017)
5. S. Li, L. Da Xu, S. Zhao, The internet of things: a survey. Inf. Syst. Front. **17**(2), 243–259 (2015)
6. I. Lee, K. Lee, The Internet of Things (IoT): applications, investments, and challenges for enterprises. Bus. Horiz. **58**(4), 431–440 (2015)
7. http://www.cisco.com
8. P. Oghazi et al., RFID and ERP systems in supply chain management. Eur. J. Manage. Bus. Econ. (2018)
9. A. ur Rehman et al., IoT and smart manufacturing, in *E-Manufacturing and E-Service Strategies in Contemporary Organizations*, (IGI Global, 2018), pp. 1–38
10. B.H. Al Lawati, X. Fang, *Home automation internet of things: adopted or diffused?*, in *International Conference on Distributed, Ambient, and Pervasive Interactions* (Springer, 2018)
11. C.-C. Chen, RFID-based intelligent shopping environment: a comprehensive evaluation framework with neural computing approach. Neural Comput. Appl. **25**(7–8), 1685–1697 (2014)

12. F. Bibi et al., A review: RFID technology having sensing aptitudes for food industry and their contribution to tracking and monitoring of food products. Trends Food Sci. Technol. **62**, 91–103 (2017)
13. A.M. Turri, R.J. Smith, S.W. Kopp, Privacy and RFID technology: a review of regulatory efforts. J. Consum. Aff. **51**(2), 329–354 (2017)
14. F. Tao et al., Advanced manufacturing systems: socialization characteristics and trends. J. Intell. Manuf. **28**(5), 1079–1094 (2017)
15. M. Gaikwad et al., *Barcode Based Student In Out System.* (2017)
16. S. Windmann et al., *A novel self-configuration method for RFID systems in industrial production environments*, in *22nd IEEE International Conference on Emerging Technologies and Factory Automation (ETFA), 2017* (IEEE, 2017)
17. Z. Irani, A. Gunasekaran, Y.K. Dwivedi, Radio frequency identification (RFID): research trends and framework. Int. J. Prod. Res. **48**(9), 2485–2511 (2010)
18. A. Azizi, A. Vatankhah Barenji, M. Hashmipour, Optimizing radio frequency identification network planning through ring probabilistic logic neurons. Adv. Mech. Eng. **8**(8), 1687814016663476 (2016)
19. http://innorfid.co.za/about/technology/
20. X. Liu et al., *Fast identification of blocked RFID tags.* IEEE Trans. Mob. Comput. (2018)
21. http://www.xyrfid.com/small-paper-uhf-rfid-label-sticker/
22. http://skyrfid.com/RFID_Label_Tag.php
23. A.K. Yetisen, *Biohacking.* Trends Biotechnol. (2018)
24. http://www.yourdictionary.com/rfid-tag
25. https://www.rfidexpert.ru/en/story/818
26. J. Zheng et al., Multiple-port reader antenna with three modes for UHF RFID applications. Electron. Lett. **54**(5), 264–266 (2018)
27. J. Zhang et al., *Feature extraction for robust crack monitoring using passive wireless RFID Antenna Sensors.* IEEE Sens. J. (2018)
28. Y. Lili et al., *A PSO clustering based RFID middleware*, in *2018 4th International Conference on Control, Automation and Robotics (ICCAR)* (IEEE, 2018)
29. Y. Lee, J. Cho, *RFID-based sensing system for context information management using P2P network architecture.* Peer-to-Peer Networking Appl. 1–9 (2018)

Chapter 3
RFID Network Planning

3.1 Overview

With advances in technology, modern manufacturing industry is bounded with data interconnectivity. The key of being successful in the modern manufacturing industry is achieving to be useful in time data which can be part of supply chain management tracking devices which are used to track and map a live subject or parts of the flexible manufacturing industry to give alerts to manufacturers in different regards to need for maintenance of parts. Manufacturers to collect required data for further analyses deploy different types of gadgets of IoT such as sensors, network cameras or most smart one RFID systems in their operational field [1].

An RFID system as discussed in the previous chapter includes four major elements: tags, readers, antennas, and computer unit. It should be reminded that the antenna as internal antenna can be a part of the reader, or it can be mounted as external antenna, and also each reader can adopt one or more than one antenna. Readers through antennas collect the data sent by tags and relay this to a host computer to process for further implementations. In essence, RFID is the technology of data processing which the data is the radio frequency signals emitted from mounted tags on subjects and sent by readers to a host computer unit (Fig. 3.1).

There are three types of tags: passive, semi-passive, and active. Differences between these are in their source of power where active and semi-passive tags are battery powered but passive tags do not have internal power. It should be noted that in this book, passive tags have been chosen to be deployed in RFID network. The reason for adopting passive tags here is because of their advantages such as being cost-effective and having a long life cycle compared to the semi-passive and active tags [3].

Data of a tag can be read by a reader in certain distance between tag and antenna of the reader. It means that a reader through its antenna can receive information of a tag in a limited range. The establishment of communication between the antenna and tag relies on the distance between tag and antenna and is highly sensitive in the

© The Author(s), under exclusive licence to Springer Nature Singapore Pte Ltd. 2019
A. Azizi, *Applications of Artificial Intelligence Techniques in Industry 4.0*,
SpringerBriefs in Applied Sciences and Technology,
https://doi.org/10.1007/978-981-13-2640-0_3

case of changing the distance, so it means that the antenna is not able to collect the data of the tags which are out of the its interrogation range. This issue can be named as uncovered tag problem.

Utilizing more antennas is a common solution which is adopted in recent years to overcome this problem [4]. By adopting more antennas in RFID system, important criteria such as number of antennas and positions of them, collision of the antennas, and coverage of the network needs to be calculated. Answering such questions has led to an important concept known as RFID Network Planning (RNP).

3.2 Mathematical Modeling

To model an RFID network mathematically, some important criteria such as number of tags, number of antennas, coverage percentage of the network, collision percentage of antennas, and transmitted power in the network should be considered in the mathematical model. One of the reliable mathematical models which are adopted by many researchers to deal with RNP is the Friis transmission equation [5]. In this book, a developed model of this equation which is proposed by Gong at el. [6] is utilized to deal with RNP.

To model an RFID network and have an RNP, the first step is defining the working area, the next step is defining number of tags which needs to be read by readers, and finally defining number of antennas of the readers (see Fig. 3.2).

An example of an RFID network plan is illustrated in Fig. 3.3 with the following specifications:

- Working area: square room with dimension of 100×100 m^2
- RFID tags: 20 passive tags which are randomly distributed in working area
- Antenna: 20 antennas which are randomly distributed in working area.

After defining the RNP, the next step is defining and calculating parameters of the RNP such as coverage of the network, redundancy of the network, interference of the network, and finally transmitted in the network.

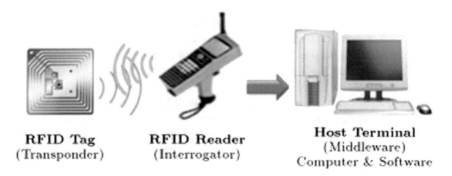

RFID Tag   **RFID Reader**   **Host Terminal**
(Transponder)   (Interrogator)   (Middleware)
                                               Computer & Software

Fig. 3.1 RFID technology components [2]

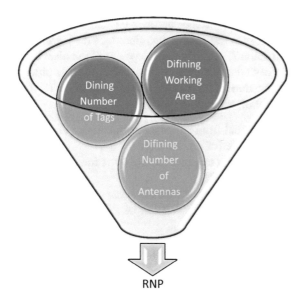

Fig. 3.2 Steps of defining an RNP

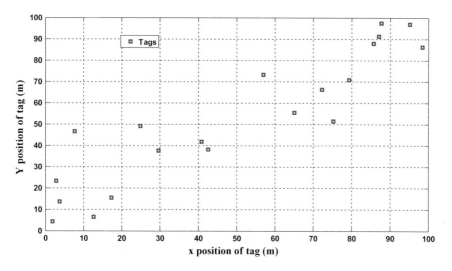

Fig. 3.3 Working area and distributed 20 tags

3.2.1 Coverage

Coverage of the network is defined as the percentage of the tags which have been read by the readers via antennas of the readers and the most important criteria which should be satisfied in RNP is achieving 100% coverage of tags.

As mentioned before, the distance between tag and antenna plays a major role in tag coverage. In the case of the passive tags since they do not have a power source to comminucate with readers, and they use the power of the recevied RF signals from the antennas to activate their internal circuts, so the distance criterion has a greater importance. The procedure of calculating the coverage of an RFID network can be summarized in the following three steps [7]:

(1) A tag receives a signal from an antenna with power of $PT_{a,t}$ which is greater than the threshold value of power of the tag (T_t).
(2) This signal activates the tag, and the tag starts to send a backscatter signal to antennas with power of $PA_{t,a}$.
(3) If the power of the backscatter signal is greater than the threshold value of the power of the antenna (T_a), then it can be said that the tag is covered by the antenna (see Fig. 3.4).

The mathematical model of the above steps is defined as follows [7]:

$$C_v(t) = \begin{cases} 1, & \text{if } \exists a_1, a_2 \in AS, \, PT_{a_1,t} \geq T_t \wedge PA_{t,a_2} \geq T_a \\ 0, & \text{otherwise} \end{cases} \quad (2.1)$$

The adopted formula for N tags is defined below [7]:

$$\text{COV} = \sum_{t \in TS} \frac{C_v(t)}{N_t} \times 100\% \quad (2.2)$$

where $N_t = 20$ and $1 \leq t \leq 20$.

Received power by tags $PT_{a,t}$ and transmitted power by tags $PA_{t,a}$ are calculated through Friis transmission equation as given below [3]:

$$PT_{a,t}[\text{dBm}] = P_1[\text{dB}m] + G_a[\text{dB}i] + G_t[\text{dB}i] - L[\text{dB}] \quad (2.3)$$

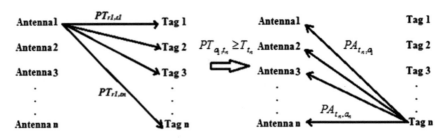

Fig. 3.4 Example of tag coverage, if $PT_{a_1,t_n} \geq T_{t_n} \wedge PA_{t_n,a_3} \geq T_{a_3}$, then $C_v(1) = 1$ [3]

$$L[\text{dB}] = 10 \log \left[\left(\frac{4\pi}{\lambda} \right)^2 d^n \right] + \delta[\text{dB}] \tag{2.4}$$

$$P A_{t,a}[\text{dBm}] = P_b[\text{dBm}] + G_t[\text{dB}i] + G_a[\text{dB}i] - 20 \log \left(\frac{4\pi d}{\lambda} \right) \tag{2.5}$$

$$P_b = (\Gamma_{\text{tag}})^2 \times p_t \tag{2.6}$$

where P_1 is the transmitted power of the antenna, G_a is the gain of the antenna, G_t is the gain of the tag, L represents loss, λ is the wavelength, d is the distance between tag and antenna, n depends on environment which varies from 1.5 to 4, δ represents other losses, and P_b is the backscatter power transmitted by tag which is reduced by multiplying into the reflection coefficient Γ_{tag} [3].

3.2.2 Redundant Antennas

Redundant antenna can be defined as an antenna which do not have an effect on coverage percentage of the network. Having redundant antenna in the network means imposing more unnecessary extra cost to the manufacturer, so it is an important criterion in RFID network planning to calculate the redundancy of the network and eliminate or reduce it. Following procedure results in calculating the number of useful antennas. The mathematical representation of this concept is shown below [3]:

$$N_a = N_{\max} - N_{\text{red}} \tag{2.7}$$

where $N_{\max}$ is the maximum number of antennas that can be deployed in the network. N_{red} is the number of redundant antennas and N_a is the efficient and useful antennas.

To calculate the number of redundant antennas, each of the deployed antennas in the network should be turned off one by one and in each step the coverage of the network in the absence of that antenna should be calculated. If the calculated coverage remains as before so it means that antenna is redundant and number of the redundant antennas should be added by one. By following the procedure for all antennas, at the end of it total number of the redundent antennas can be calculated.

3.2.3 Interference

Interference is another important criterion in RNP which can be defined as collision between the efficient antennas. Collision is a kind of interference in the network and is a resultant of interrogating one tag by more than one antenna of readers.

Below equation is a mathematical representation of a collision which is accrued by one tag [7]:

$$\gamma(t) = \sum PT_{a,t} - \max\{PT_{a,t}\}, a \in AS \wedge PT_{a,t} \geq T_t \qquad (2.8)$$

And the total interference of the network can be calculated by sum of all collisions of the network. The mathematical representation of total interference of the network is shown below as [7]:

$$\text{ITF} = \sum_{t \in TS} \gamma(t) \qquad (2.9)$$

3.2.4   Transmitted Power

As noted before, the distance between tag and antennas have a major role in network coverage. Since the interrogation range of each antenna has a direct relation with its transmitted power, so reducing the amount of transmitted power may cause in decreasing coverage of the network. It means that however reducing the total transmitted power is one of the RNP goals but it has the lowest criteria in network planning [3].

$$\text{POW} = \sum_{a \in AS} PS_a \qquad (2.10)$$

where PS_a is the amount of transmitted power by antennas.

After the introduction to RFID network planning at the beginning of this chapter, criteria which are important and should be calculated in RNP have been introduced. It should be noted that RNP mathematical representation is not capable of making a change in coverage, interference, or redundancy of the network by changing the number of deployed antennas, so in the next chapter, different RNP optimization processes which can be addressed to satisfy these criteria will be investigated and proposed.

References

1. Y. Bai, *Industrial Internet of Things over tactile internet in the context of intelligent manufacturing.* Cluster Comput. 1–9 (2017)
2. J. Majrouhi Sardroud, Influence of RFID technology on automated management of construction materials and components. Sci. **19**(3), 381–392 (2012)
3. A. Azizi, A. Vatankhah Barenji, M. Hashmipour, Optimizing radio frequency identification network planning through ring probabilistic logic neurons. Adv. Mech. Eng. **8**(8), 1687814016663476 (2016)

4. L. Hervert-Escobar et al., Optimal location of RFID reader antennas in a three dimensional space. Ann. Oper. Res. **258**(2), 815–823 (2017)
5. O. Franek, Phasor alternatives to Friis' transmission equation. IEEE Antennas Wirel. Propag. Lett. **17**(1), 90–93 (2018)
6. Y.-J. Gong et al., Optimizing RFID network planning by using a particle swarm optimization algorithm with redundant reader elimination. IEEE Trans. Industr. Inf. **8**(4), 900–912 (2012)
7. A. Azizi, *Introducing a novel hybrid artificial intelligence algorithm to optimize network of industrial applications in modern manufacturing.* Complexity **2017** (2017)

Chapter 4
Hybrid Artificial Intelligence Optimization Technique

4.1 Overview

RFID technology as a gadget of IoT has been utilized in modern manufacturing to enable manufacturers to track and identify objects or parts to get the required data. Fulfillment of this purpose needs to equip objects with RFID tags and utilizes RFID antennas in certain places to enable readers to collect data of the objects. Some criteria such as the collision of these antennas, the coverage of network, and transmitted power in the network are calculated through a mathematical model. Calculating these criteria and calculating the number of required antennas for RFID network lead to concept of RFID Network Planning (RNP) and in the higher level concept of optimizing RNP.

Conventional method which has been used in the past is trial-and-error approach, which was not an efficient and accurate solution to optimize RNP and could only be implemented to not large-scale network. In recent years with advances in technology and software engineering modern artificial intelligence computational techniques which are more efficient than conventional trial-and-error technique has been started to utilize to deal with optimizing RNP.

The main idea behind artificial computational intelligent methods is that they are not concentrating on one solution; on contrary, these methods start with initial solution and continue with searching the best solution among all possible solutions. It means that in each iteration the best solution should be selected, and it continues till in the final iteration the best solution among of the best possible solutions be selected.

Artificial computational intelligent techniques such as Artificial Neural Networks [1], Fuzzy Logic [2], Genetic Algorithms (GA) [3–5], Particle Swarm Optimization (PSO) [6–8], Differential Evolution (DE) [9], and hierarchical artificial bee colony algorithm [10] are a point of interest for many scientists working with the RNP problem. In this respect, Feng and Qi [11] for solving and optimizing complicated RNP problems proposed a novel optimization algorithm which is a combination

© The Author(s), under exclusive licence to Springer Nature Singapore Pte Ltd. 2019   27
A. Azizi, *Applications of Artificial Intelligence Techniques in Industry 4.0*,
SpringerBriefs in Applied Sciences and Technology,
https://doi.org/10.1007/978-981-13-2640-0_4

of genetic algorithm and particle swarm optimization techniques which is known as multi-community GA-PSO. PS^2O optimization algorithm has been adopted by Chen et al. [12] by concentrating on optimizing position of the deployed antennas of readers in the network. To optimize RNP parameters, PSO-based solution has been proposed by Nawawi et al. [13]. Lu and Yu [14] has concentrated on optimizing coverage of RNP by multidimensional optimizing k-coverage model, and finally one of the most recent researches which concentrate on combining two optimization techniques has been conducted by Gong et al. [8]; the mentioned research has combined PSO algorithm with redundant reader elimination for optimizing RNP.

Most of the studies which have been reviewed in this book have concentrated on satisfying RNP criteria based on the optimizing positions of deployed antennas, and these researches do not deal with optimizing number of the antennas. In this case, the number of the antennas remains constant and just position of them is changed by optimization techniques.

It has a great importance to optimize all RNP criteria, because the goal of optimizing RFID network is not only optimizing criteria such as coverage, interference, etc. of the network but also calculating the number of antennas in a cost-efficient manner that should be considered as one of the main criteria. In brief, the goal of RNP and optimizing it can be summarized in the following statement: To plan a cost-efficient RFID network, it is necessary to minimize the number of antennas, minimize interference of antennas, and maximize coverage area of objects [15].

Therefore, in this book, to satisfy these targets and design an efficient RFID network, a hybrid optimization technique is introduced. The Hybrid term refers to combining two different algorithms to optimize RNP as one optimization approach (see Fig. 4.1).

In this purpose, Ring Probabilistic Logic Neural Network (RPLNN) as a novel technique in RNP optimization is introduced. The algorithm of RPLNN is designed in a way which has been made capable of the paradigm to adjust the number of embedded antennas in the network, so that it makes RPLNN optimization technique an efficient artificial computational intelligent optimization approach to deal with complex RFID network planning problems.

Fig. 4.1 Proposed hybrid optimization technique

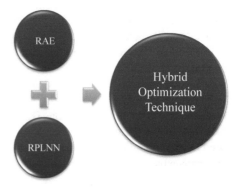

Fig. 4.2 RNP optimization process

The second component of the proposed hybrid algorithm is utilizing Redundant Antenna Elimination (RAE) optimization technique in addition to RPLNN optimization technique. Utilizing RAE, algorithm has two advantages: the first privilege is reducing optimization process by reduction of iterations and the other one is giving flexibility to RNP in terms of number of antennas.

The priority of the combined algorithms in the proposed hybrid optimization process of RNP belongs to RAE paradigm, and it has been used before RPLNN technique. It means that first RAE eliminates redundant antennas in the network in each step of optimization and then the RPLNN optimization technique will be applied to only nonredundant antennas (see Fig. 4.2).

After giving an introduction about the goal of optimization of RNP and criteria which should be satisfied by optimizing RNP, different computational artificial intelligence optimization techniques have been reviewed and a hybrid artificial intelligence optimization algorithm has been proposed to deal with the RNP. The remaining sections of this chapter have been organized as follows: first, the methodology of this research which is the proposed hybrid artificial intelligence algorithm and its components has been introduced and discussed in detail, to have a comparison between the proposed approach and other existed approaches, and in the second part, another artificial intelligent technique, namely, genetic algorithms as a well-known optimization technique in evolutionary optimization approaches, has been introduced. In the next chapter, Chap. 5, both of the approaches are implemented to optimize an RNP and results are compared.

4.2 Methodology

The aim of this book is to introduce a novel hybrid artificial intelligence optimization technique which is capable to deal with complex RNP problems. Proposed hybrid paradigm is capable to adjust the number of deployed antennas in the RFID network as same as optimizing other criteria of RNP such as achieving maximum tag coverage in the network, minimizing antenna collision, and interference of the network. Proposed hybrid algorithms have been made of two artificial intelligence optimization paradigms which perform optimization process in the form of a series network. The first phase of the optimization has been done through Redundant Antenna Elimination (RAE) algorithms and it has been followed by Ring Probabilistic Logic Neural Network (RPLNN) paradigm to accomplish the optimization process (see Fig. 4.2). Flowchart of the proposed hybrid artificial intelligence optimization paradigm has been shown in Fig. 4.3.

The first step of the optimization process of RNP starts with solving the proposed mathematical model of RNP which is discussed in the previous chapter of this book (see Fig. 4.3). All parameters of RNP which are coverage, redundancy, and interference of the RFID network should be calculated. To fulfill this task, the number of deployed RFID antennas in the network should be known, so in the first iteration, this number has an arbitrary value. After calculating all parameters of RNP, the next step is optimization process through Redundant Antenna Elimination (RAE) algorithm. In this step of proposed hybrid artificial intelligence algorithm, all antennas which deployed as non-efficient antennas and impose extra cost to network which is the resultant of redundancy are eliminated and deleted from RFID network. After the RAE optimization process has been finalized, it is the time to calculate the total number of nonredundant antennas which remain on the RFID network. The next step of optimization process is to apply the second phase of hybrid artificial intelligence algorithm which is Ring Probabilistic Logic Neural Network (RPLNN) algorithm. It is important to know that RPLNN algorithm optimizes the RNP based on nonredundant antennas, which has been calculated by RAE algorithm. The next step is to test if optimization process is finished or not, and to perform this task, the number of iterations which are required for whole hybrid optimization process to optimize RNP should be defined by the user. After the phase of optimizing RNP utilizing RPLNN technique, if iterations have reached to predefined number of iterations, in this case, optimization process should be stopped and best solution for RNP should be chosen. If iterations have not been completed, in this case, maximum number of deployed antennas in the network should be calculated and RNP optimization process from the step of solving the mathematical model for RNP should be repeated. The hybrid artificial intelligence optimization process continues till iterations reach predefined number of iterations.

After an introduction of proposed hybrid artificial intelligence algorithms and overviewing the flowchart of the algorithm, the whole optimization process has been introduced. In the next section, the two components of hybrid artificial intelligence algorithm which has been adopted to deal with RNP problem have been introduced. These algorithms which are known as redundant antenna elimination algorithm and ring probabilistic logic neural networks in the next sections have been discussed in detail.

4.2.1 Redundant Antenna Elimination Algorithm

The logic of the optimization process through Redundant Antenna Elimination (RAE) paradigm is based on deleting redundancy on the RFID network. Terms of redundancy mean that an existed RFID tag has been covered by two different RFID antennas of

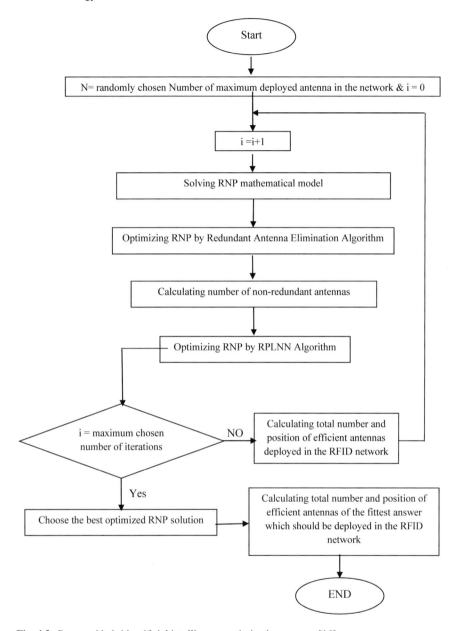

Fig. 4.3 Proposed hybrid artificial intelligence optimization process [16]

readers. It happens when a tag receives an activation signal from an antenna and starts to backscatter this signal; the emitted signal from the tag can be received by more than one antenna and tag will be covered by more than one antenna. Since coverage of the network by eliminating all of the antennas except one of them does not be changed and remains constant, it means that the other antennas are considered as redundant antennas on the network and should be eliminated.

In the first step of the optimization process, RAE algorithm optimizes RNP in each iteration of the optimization task by eliminating redundant deployed antennas by performing the following procedure.

First, coverage of the RFID network by the RNP mathematical model which has been introduced previously has been calculated; the next step is to calculate the redundancy of the network by the mathematical formulation which has been discussed in Sect. 3.2.2 of this study. To perform this task, deployed antennas in the RFID network should be eliminated one by one and coverage of the RFID network should be calculated after each antenna elimination. If the total calculated coverage of the RFID network after each antenna elimination remains as before antenna elimination, then the eliminated antenna remains as deleted; otherwise, if the total calculated coverage of the network after the antenna elimination be less than the RFID network coverage before the antenna elimination, then eliminated antenna should be recovered and undeleted.

This procedure should be repeated for all the antennas; each calculation step has been performed for each antenna known as an iteration, so the number of the iterations of this algorithm should be equal to total number of deployed antennas into the RFID network.

In each iteration of the optimization utilizing RAE paradigm if a redundant antenna has been founded by the algorithm, in this case, the number of the redundant antennas which has been eliminated will be added by one.

After performing the last iteration of the RAE optimization process which is investigation redundancy of the last deployed antenna in the RFID network, total number of redundant antenna can be calculated and by calculating this number and knowing the total deployed antenna in the RFID network, the total of nonredundant antennas has been calculated and has been prepared to go for next step of the hybrid optimization technique which is performing optimization through RPLNN paradigm on nonredundant antennas.

The proposed RAE optimization algorithm can be summarized in the following steps (see Fig. 4.4):

a. Total coverage of RFID network has been calculated (C_1).
b. First, deployed antenna in the RFID network has been eliminated.
c. After the antenna elimination, again total coverage of RFID network has been calculated (C_2).
d. If coverage of the network after antenna elimination has been calculated less than before antenna elimination ($C_2 < C_1$), then eliminated antennas must be recovered.
e. If coverage of the network after antenna elimination has been calculated equal to before antenna elimination ($C_2 = C_1$), then the antenna should remain eliminated and the number of redundant antennas has been added by one.

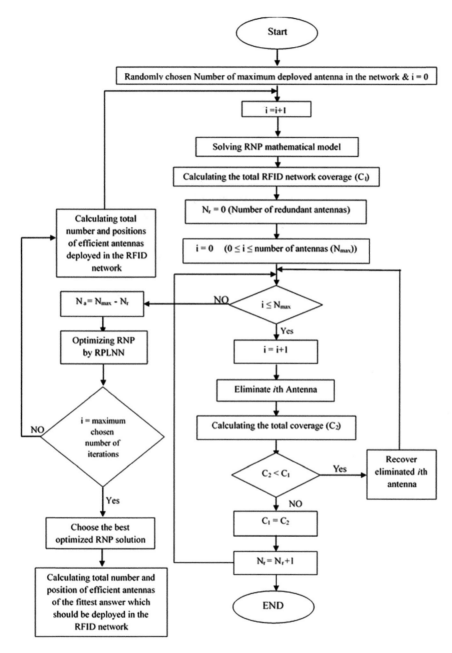

Fig. 4.4 Flowchart of the proposed redundant antenna elimination algorithm [16]

f. The RAE optimization process has been ended after investigating the redundancy status of deployed antenna in the network, and by then, the number of nonredundant antennas has been calculated and has been ready for next phase of hybrid artificial intelligence optimization process.

4.2.2 Ring Probabilistic Logic Neural Networks

The second part of the proposed hybrid optimization technique is Ring Probabilistic Logic Neural Network (RPLNN) which also is known as Ring Probabilistic Logic Node (RPLN). RPLNN paradigm is a part of RAM-based Weightless Artificial Neural Network (WANN), so before proposing the utilized structure of this algorithm in this book a brief review of Neural Network (NN), RAM-based WANN, structure of a Probabilistic Logic Neuron (PLN), and RPLNN have been given in the following sections.

4.2.2.1 Neural Networks

Networks of biological neurons which are connected to central nervous system to perform a specific physiological function conventionally have been known as Neural Network (NN) (see Fig. 4.5), but in past decades with advances in software engineering a new term which relies on artificial neurons has been proposed and known as Artificial Neural Network (ANN) [17] (see Fig. 4.6).

As shown in Fig. 4.5, to build a biological neural networks more than one neuron can be connected to a single neuron through axons and dendrites. Through the connection points which are known as synapses, neurons communicate with central nerve system by sending electrical signals [18].

The purpose of artificial neural network models is to simulate functions of biological neural networks. ANN has been utilized different engineering and science fields such as control [19], data processing [20], robotics [21], function approximation [22], pattern, and speech recognition [23].

As shown in Fig. 4.6, ANN has been made of interconnecting neurons which have been categorized into three layers which are input layer, hidden layer, and output layer. Based on the type of connections between these neurons, artificial neural networks can be divided into two different groups: Weighted artificial neural networks and weightless artificial neural networks [16].

After a brief introduction about neural networks, since RPLNN has been categorized as part of weightless artificial neural networks, the next section has been organized to give an overview of weightless artificial neural networks.

Weightless Artificial Neural Networks

Weightless artificial neural networks have been known as simply implemented artificial intelligent dynamic paradigms which had been proposed for the first time in

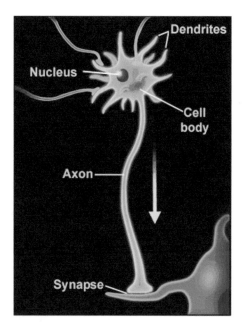

Fig. 4.5 Biological neuron [18]

Fig. 4.6 Example of
artificial neural networks
[18]

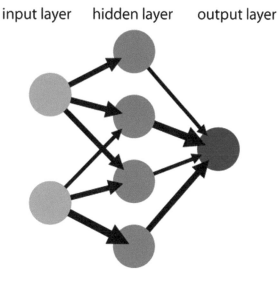

1965 by Aleksander et al. [24]. The proposed idea was based on design devices
which have random access memories. Bledsoe and Browning [25] were the first
who implemented the idea of weightless artificial neural networks to build pattern
recognition device. General Neural Unit model (GNU) which was a novel model

in weightless artificial neural networks modeling was proposed by Aleksander et al. [26]. The proposed model was a building block to build cognitive structures. In 1993, Aleksander and Morton [27] proved that the machine state artificial consciousness can be discussed by weightless artificial neural network models. It means that complex cognitive behavior of humans can be modeled by utilizing such machines as building blocks.

The idea of utilizing Random Access Memory (RAM) is the base of weightless artificial neural networks. In general, RAM refers to a device with more than one input and just one output, or in the other word RAM is a multi-input single-output (MISO) device. Inputs and outputs should be in binary form, and it means that based on the input data which is in the form of 0 or 1 various locations of memory can be entered (see Fig. 4.7).

Since the inputs are in binary form and each set of the inputs can have access to only one single location of stored output, then for N inputs number of the output locations in memory should be 2^N.

RAM Neural Networks

More than one RAM neuron can be utilized to build up a RAM network. As shown in Fig. 4.8, RAM network consists of two layers namely RAM layer and output layer. RAM layer is built of K different RAM neurons in which each of them has input vector with size of N. The output layer is nothing more than a summation node in which output values of all RAM neurons are added up in that node. It should be noted that in concept of RAM networks length of each set of the input data is one bit [29].

The training algorithm of a RAM neural network is described as below [30]:

(1) Start
(2) Giving the input in form of binary code (0 or 1) which is known as pattern should be followed to first RAM.
(3) If the input can access the memory location, output value should be set up to 1 $(F = 1)$.
(4) If the input cannot access the memory location, output value should be set up to 0 $(F = 0)$,

Fig. 4.7 A RAM neuron [28]

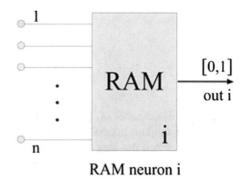

RAM neuron i

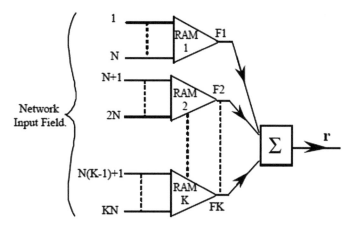

Fig. 4.8 A RAM discriminator [30]

(5) Repeat steps 1–3 for all RAM neurons.
(6) Calculate sum of all outputs ($\sum F = r$).
(7) If calculated sum of all outputs is equal to total number of RAMs, then pattern is recognized ($r = k$).
(8) If calculated sum of all outputs is less than total number of RAMs, then pattern is not or partially recognized ($r < k$).

The discussed training algorithm of RAM neural network is summarized as a paradigm flowchart in Fig. 4.9.

After giving a brief introduction with regard to weightless neural networks and idea of constructing and developing this concept based on random access memory which is known as RAM, the RAM neural network and the training paradigm of this network have been introduced and discussed. Following after, in the next section, ring probabilistic logic neuron as advanced concept in building weightless neural models has been introduced.

Probabilistic Logic Neural Networks

The Probabilistic Logic Neuron (PLN) is a RAM-based device which is proposed in 1988 by Aleksander et al. [4]. As shown in Fig. 4.10, a PLN is made up of a probabilistic node which based on the input array calculates the output array. Probabilistic node term refers to the probability percentage of each node which can be calculated by dividing the output of a PLN to maximum output and can be stored in memory.

PLNs from point of the value and length of the output have two major differences with RAM neurons. RAM neurons can store their output in length of one bit in memory location; however, PLNs are capable of store outputs which have greater length than one bit which is known as B-bit. The other difference is that the outputs of RAM neurons can be 0 or 1; however, outputs of PLNs rather than having these two

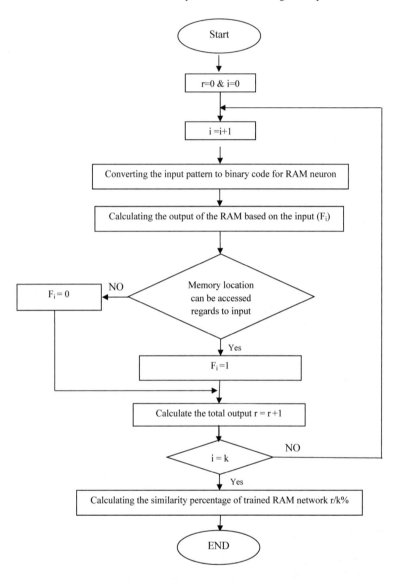

Fig. 4.9 Flowchart of training algorithm of RAM neural network

options have a third state which is "Don't Care" state and represented by "*X*". "Don't Care" state refers to the outputs which can have value of 0 or 1 with probability of 50% [16].

By adopting the "Don't Care" state, the meaning of the output 0 has been changed in PLNs. It means that state of 0 has two meanings in PLN concept: first, it can be interpreted that for the given input vector the PLN neuron has not been trained, and

Fig. 4.10 Probabilistic
Logic Neuron (PLN) [16]

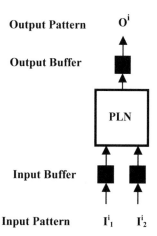

second, it can be interpreted as the calculated output of the trained PLN which is opposite of the RAM neuron concept. It can be said that utilizing this third state enabled PLNs to have better performance compared to RAM neurons, and as an important option they are opposite to RAM; they can have more than two layers. As shown in Fig. 4.11, PLNs can be combined and build a PLN neural network with paramedical structure.

The PLN neural network can deal with hard learning problems and consists of many PLN neurons (shown as size of W) in which each has N inputs and more

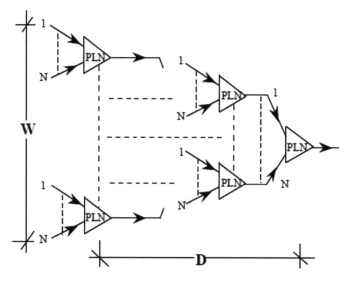

Fig. 4.11 Pyramidal PLN neural network structure [16]

than one hidden layer (shown as size of D) which ensures robustness of PLN neural networks.

In general, PLN neural networks do not have a rigid structure and based on the requirements of the problem by combining PLNs with different manners, different PLN neural networks can be designed (see Figs. 4.11 and 4.12).

These kinds of PLN neural networks also are known as Multilayer PLN network (MPLN). The efficiency of MPLNs network has been investigated and analyzed in detail by Zeng et al. [31]. The research result indicates that being flexible in arrangements of PLNs gives an advantage of having fast and efficient convergence time to find the final solution of hard learning problems by MPLNs.

Ring Probabilistic Logic Neural Networks

In the year 2002 based on the flexibility of structural design of MPLNs, a new structure which is known as Ring Probabilistic Logic Neural Networks (RPLNN) to deal with optimization problems is proposed by Menhaj and Seifipour [32]. The proposed structure connects input of the first PLN to output of the last PLN, so this connection based on the available datasets can be divided into two different groups which are feedforward and feedback connections.

The feedforward connection state happens when input data of the first PLN be available, so that this data goes as feedforward signal to the output of the last PLN. The feedback connection state happens when the output data of the last PLN be available, so that this data goes as feedback signal to the input of the first PLN. This feedback and feedforward connection which connects first and last PLNs is known as ring structure, and because of this property, the name of this kind of MPLN network is known as Ring Probabilistic Logic Neural Network (RPLNN). In 2016 Azizi et al. [33]

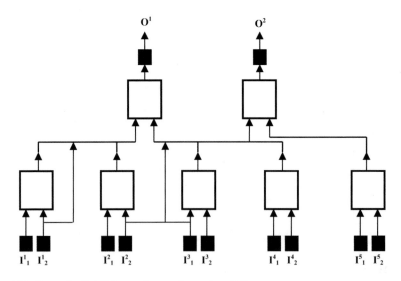

Fig. 4.12 Example of a PLN neural network structure [16]

utilized RPLNN structure as a part of weightless neural networks to optimize weighted artificial neural network model of mechanical behavior of friction steer welding (see Fig. 4.13). A special structure of this algorithm from point of defining inputs and outputs of the RPLNN has been implemented in this book to deal with RNP problem which has been discussed in detail in the next chapter of this book.

In RPLNN structure, each PLN has its own truth table which sets the output value based on the input vector to 0, 1 or don't care states. The training of RPLNN can be summarized as following: the training process continues till all don't care states have been replaced with values 0 or 1 [34]. It means that RPLNN structure as the second phase of proposed hybrid artificial intelligence optimization paradigm fulfills the task of optimization based on pure random search among all possible solutions.

The RPLNN optimization process starts with converting the inputs in the form of binary codes which is made of zeroes and ones; it is followed by creating a population of possible answers to the problem. The next step is to evaluate each of the existed answers in the defined population to know which of them is the best answer and has the fittest value; so to perform the evaluation task, defining a fitness function is essential. It should be noted that the fitness function should be defined per criteria of the problem which requires to be optimized and differs from one problem to another. The fitness function which has been used in this book to optimize RNP has been introduced in detail in the next chapter.

The proposed RPLNN algorithms which have been adopted in this book as a part of the proposed hybrid artificial intelligent algorithm are shown as flowchart in Fig. 4.14.

After calculating the total number of nonredundant antennas by the first part of proposed hybrid optimization paradigm which is RAE, the second phase of optimization is started with RPLNN by calculating all criteria of RNP mathematical model and by utilizing positions of nonredundant antennas. The calculated positions of nonredundant antennas at the parallel procedure are encoded as binary code to create the population of all possible answers. The detailed discussion about how to

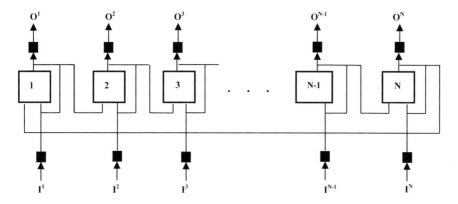

Fig. 4.13 RPLNN structure [16]

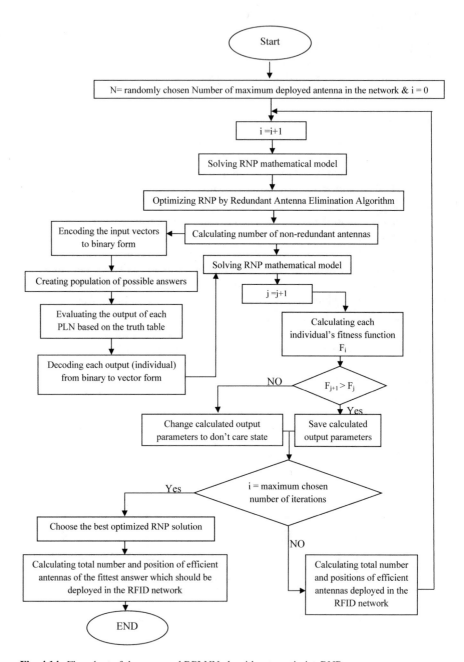

Fig. 4.14 Flowchart of the proposed RPLNN algorithm to optimize RNP

create the population of all possible answers has been given in the next chapter. Each existed answer in population is known as individual, and the next step is to evaluate each of these individuals by defined fitness function to optimize RNP which has been discussed in detail in the next chapter. In the same time, parallel process output of each PLN in the proposed RPLNN structure should be calculated by PLN truth table with regard to encoded binary positions of nonredundant antennas as inputs. The next step is to decode the calculated binary output of decimal vector which is new position of antennas and solve RNP mathematical model based on these new positions of antennas. The next step is to calculate fitness function of RNP based on calculated criteria which is achieved by applying new positions of antennas. The next step is to compare two calculated fitness functions based on two different positions of antennas given as RPLNN inputs and outputs; if the fitness function of RPLNN output be less than the calculated fitness function of RPLNN input, then all calculated RPLNN outputs which are the binary forms of positions of antennas should be returned back to don't care status; otherwise, all calculated outputs should be saved as calculated. These processes continue till the optimization process completes all predefined number of iterations, and the last step is to choose the best answer of the calculated population of the antennas based on its fitness function. It means that the fittest individual of the population of answers is the best solution of the proposed optimization algorithm.

The second phase of proposed hybrid artificial intelligence algorithms, RPLNN, which has been adopted to deal with RNP in this book can be summarized in the following steps:

1. A. Encode the positions of nonredundant antennas calculated by RAE to binary codes and create population of possible answers.
   B. Solve RNP for calculated nonredundant positions of antennas by RAE.
2. A. Calculate the output of each PLN of RPLNN by truth table of PLN with regard to inputs.
   B. Decode RPLNN outputs to decimal position of antennas vectors.
   C. Solve RNP for calculated positions of antennas.
3. A. Calculate the fitness function value for each individual of inputs of RPLNN which are calculated positions of antennas.
   B. Calculate the fitness function value for each individual of outputs of RPLNN which are calculated positions of antennas.
4. A. If the calculated fitness function value of output of RPLNN be more than the calculated input fitness function value, then save the value of the outputs of the RPLNN.
   B. If the calculated fitness function value of output of RPLNN be less than or equal to the calculated input fitness function value, then reset the value of the outputs of the RPLNN and set all of them as don't care state.
   C. Repeat these steps for the predefined number of iterations.
5. Choose the best and fittest answer between all possible solutions.
6. Calculate the number and positions of efficient antennas which should be deployed on RFID network.

After introducing the components of the proposed hybrid algorithm, in the next section to have a comparison of the performance of the proposed algorithm and other algorithms which have been adopted by other researchers, a well-known evolutionary optimization technique known as Genetic Algorithm (GA) has been introduced to optimize the RNP.

It has a great importance to know in this book that GA has not been adopted as sole optimization technique and it has been combined by RAE.

4.3   Genetic Algorithm

Genetic algorithm is one of the well-known evolutionary optimization techniques which has been adopted by many researches to optimize complex problems [35, 36]. Briefly, the optimization process by GA can be divided into six steps as follows [37]:

(1)   Creating population of possible answers,
(2)   Evaluating fitness function,
(3)   Creating next generation of possible answers,
(4)   Applying crossover,
(5)   Applying mutation, and
(6)   Repeat steps 2–5.

The first and the second steps which are creating population of possible answers and evaluating performance of them by fitness function have the same procedure as have been discussed in the implementation chapter of this book.

The next step is to create the next generation of the population of possible answers by adopting an appropriate selection procedure. In this book, roulette wheel selection approach has been utilized to select the best answer through calculating the fractional fitness function of each possible answer which has been defined below [37]:

$$F(x_i) = \frac{f(x_i)}{\sum_{i=1}^{n} f(x_i)} \quad i = 1 \dots \text{number of antennas} \tag{4.1}$$

According to the number of the possible answers of the population which has been taken as 100 in this book, the roulette wheel should be spinet for 100 times to select an answer to generate the next generation of the possible answers (see Fig. 4.15).

The next step is to apply crossover to the generated population. Crossover operator combines different parts of two different answers (chromosomes) known as parents and produces new chromosomes known as children. Crossover can be adopted as two-point crossover (see Fig. 4.16a) or single-point crossover (see Fig. 4.16b). In this book, single-point crossover has been adopted as the operator of GA.

The final step of optimization by GA is to adopt another operator known as mutation. The mutation operates as NAN function of one of the binary bits of the answers (gen). It means that if the gene has binary value of 0, the mutation operator

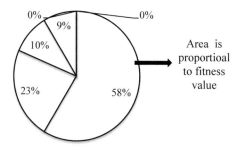

Fig. 4.15 Example of roulette wheel [37]

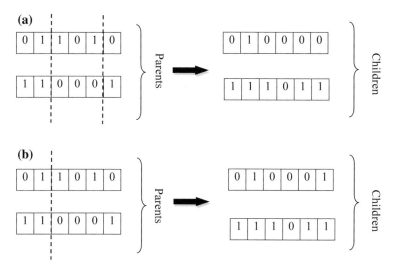

Fig. 4.16 **a** Two-point crossover, **b** single-point crossover [16]

Fig. 4.17 Example of mutation [37]

will change it to 1 and if it has binary value of 1, the operator will change it to 0 (see Fig. 4.17).

These should be repeated till predefined criteria be satisfied, in which this criterion in this book has been taken as completing 100 iterations of the optimization process.

After introducing the proposed hybrid artificial intelligence paradigm and the components of it, a well-known evolutionary optimization technique which is known as genetic algorithm has been introduced in detail. In the next chapter, both algorithms have been implemented for optimizing the proposed RNP and results have been analyzed and discussed in detail.

References

1. R. Kuo et al., The application of an artificial immune system-based back-propagation neural network with feature selection to an RFID positioning system. Robot. Comput.-Integr. Manuf. **29**(6), 431–438 (2013)
2. T. Garcia-Valverde et al., A fuzzy logic-based system for indoor localization using WiFi in ambient intelligent environments. IEEE Trans. Fuzzy Syst. **21**(4), 702–718 (2013)
3. Q. Guan et al., Genetic approach for network planning in the RFID systems, in *Sixth International Conference on Intelligent Systems Design and Applications, 2006. ISDA'06* (IEEE, 2006)
4. Y. Yang et al., A RFID network planning method based on genetic algorithm, in *International Conference on Networks Security, Wireless Communications and Trusted Computing, 2009. NSWCTC'09* (IEEE, 2009)
5. R. Sankarkumar, D.C. Ranasinghe, Watchdog: a novel, accurate and reliable method for addressing wandering-off using passive RFID tags, in *Proceedings of the 11th International Conference on Mobile and Ubiquitous Systems: Computing, Networking and Services* (Institute for Computer Sciences, Social-Informatics and Telecommunications Engineering, ICST, 2014)
6. I. Bhattacharya, U.K. Roy, Optimal placement of readers in an RFID network using particle swarm optimization. Int. J. Comput. Netw. Commun. **2**(6), 225–234 (2010)
7. H. Chen et al., Multiobjective RFID network optimization using multiobjective evolutionary and swarm intelligence approaches. Math Probl. Eng. **2014** (2014)
8. Y.-J. Gong et al., Optimizing RFID network planning by using a particle swarm optimization algorithm with redundant reader elimination. IEEE Trans. Industr. Inf. **8**(4), 900–912 (2012)
9. L. Ma et al., Cooperative artificial bee colony algorithm for multi-objective RFID network planning. J. Netw. Comput. Appl. **42**, 143–162 (2014)
10. R.V. Barenji, A.V. Barenji, M. Hashemipour, A multi-agent RFID-enabled distributed control system for a flexible manufacturing shop. Int. J. Adv. Manuf. Technol. **71**(9–12), 1773–1791 (2014)
11. H. Feng, J. Qi, Optimal RFID networks planning using a hybrid evolutionary algorithm and swarm intelligence with multi-community population structure, in *2012 14th International Conference on Advanced Communication Technology (ICACT)* (IEEE, 2012)
12. H. Chen et al., RFID network planning using a multi-swarm optimizer. J. Netw. Comput. Appl. **34**(3), 888–901 (2011)
13. A. Nawawi, K. Hasnan, S. Ahmad Bareduan, Correlation between RFID Network Planning (RNP) Parameters and Particle Swarm Optimization (PSO) Solutions, in *Applied Mechanics and Materials* (Trans Tech Publ, 2014)
14. S. Lu, S. Yu, A fuzzy k-coverage approach for RFID network planning using plant growth simulation algorithm. J. Netw. Comput. Appl. **39**, 280–291 (2014)
15. A. Azizi, A. Vatankhah Barenji, M. Hashmipour, Optimizing radio frequency identification network planning through ring probabilistic logic neurons. Adv. Mech. Eng. **8**(8), 1687814016663476 (2016)
16. A. Azizi, Introducing a novel hybrid artificial intelligence algorithm to optimize network of industrial applications in modern manufacturing. Complexity **2017** (2017)
17. A. Azizi, Designing of artificial intelligence model-free controller based on output error to control wound healing process. Biosens J **6**(147), 2 (2017)
18. A. Azizi et al., Introducing neural networks as a computational intelligent technique, in *Applied Mechanics and Materials* (Trans Tech Publ, 2014)
19. Y. Wei et al., Improved stability and stabilization results for stochastic synchronization of continuous-time semi-Markovian jump neural networks with time-varying delay. IEEE Trans. Neural Netw. Learn. Syst. **29**(6), 2488–2501 (2018)
20. D. Bacciu et al., Randomized neural networks for preference learning with physiological data. Neurocomputing **298**, 9–20 (2018)

21. S. Levine et al., Learning hand-eye coordination for robotic grasping with deep learning and large-scale data collection. Int. J. Robot. Res. **37**(4–5), 421–436 (2018)
22. A. Azizi, N. Seifipour, Modeling of dermal wound healing-remodeling phase by Neural Networks, in *International Association of Computer Science and Information Technology-Spring Conference, 2009. IACSITSC'09* (IEEE, 2009)
23. F. Grassia et al., Spike pattern recognition using artificial neuron and Spike-Timing-Dependent Plasticity implemented on a multi-core embedded platform. Artif. Life Robot. **23**(2), 200–204 (2018)
24. I. Aleksander, T. Clarke, A. Braga, Binary neural systems: combining weighted and weightless properties. Intell. Syst. Eng. **3**(4), 211–221 (1994)
25. W.W. Bledsoe, I. Browning, Pattern recognition and reading by machine, in *Papers presented at the 1–3 December 1959, eastern joint IRE-AIEE-ACM computer conference* (ACM, 1959)
26. I. Aleksander et al., *A brief introduction to Weightless Neural Systems*, in *ESANN* (Citeseer, 2009)
27. I. Aleksander, H. Morton, *Neurons and Symbols: The Stuff that Mind is Made of*, vol. 3 (Chapman & Hall, 1993)
28. C.B. Prado et al., The influence of order on a large bag of words, in *Eighth International Conference on Intelligent Systems Design and Applications, 2008. ISDA'08* (IEEE, 2008)
29. A. Graves, G. Wayne, I. Danihelka, *Neural Turing Machines.* arXiv preprint arXiv:1410.5401 (2014)
30. S. Nurmaini, A. Zarkasi, Simple pyramid RAM-based neural network architecture for localization of swarm robots. J. Inf. Process. Syst. **11**(3) (2015)
31. L. Zeng et al., Quality driven web services composition, in *Proceedings of the 12th International Conference on World Wide Web* (ACM, 2003)
32. M.B. Menhaj, N. Seifipour, Function optimization by RPLNN. Neural Netw. 1522–1527 (2002)
33. A. Azizi et al., Modeling mechanical properties of FSW thick pure copper plates and optimizing it utilizing artificial intelligence techniques. Sensor Netw Data Commun **5**(142), 2 (2016)
34. J. Austin, A review of RAM based neural networks, in *Proceedings of the Fourth International Conference on Microelectronics for Neural Networks and Fuzzy Systems, 1994* (IEEE, 1994)
35. J. Dias et al., A genetic algorithm with neural network fitness function evaluation for IMRT beam angle optimization. CEJOR **22**(3), 431–455 (2014)
36. R. Martínez-Soto, O. Castillo, J.R. Castro, Genetic algorithm optimization for type-2 non-singleton fuzzy logic controllers, in *Recent Advances on Hybrid Approaches for Designing Intelligent Systems* (Springer, 2014), pp. 3–18
37. A. Ashkzari, A. Azizi, Introducing genetic algorithm as an intelligent optimization technique, in *Applied Mechanics and Materials* (Trans Tech Publ, 2014)

Chapter 5
Implementation

5.1 Overview

Implementation of the proposed hybrid artificial intelligence algorithm to solve and optimize an RNP has three phases which are defining working area which an RFID network should be established and optimized, defining the parameters of the proposed algorithm, and implementing the optimization algorithm to defined RFID network.

5.2 Working Area

The first step of design an RFID network is defining the working area which RFID tags should be covered by deployed antennas, and the proposed hybrid artificial intelligence algorithm has to be implemented to this working area to optimize number and positions of deployed RFID antennas. To perform this task, two different working areas which have various properties have been defined in this book. The first one is the static working area, and the other one is the dynamic working area.

5.2.1 Static Working Area

The proposed static working area in this book is a square room with dimension of 100×100 m^2 and consists of 20 RFID passive tags which randomly have been distributed in this working area. The static state refers to the movement state of the tag which has been considered as not moving tag. A good example of this status can be given as parts which have been embedded with RFID tags in warehouses. The proposed static working area has been modeled with MATLAB software (see Fig. 5.1).

© The Author(s), under exclusive licence to Springer Nature Singapore Pte Ltd. 2019   49
A. Azizi, *Applications of Artificial Intelligence Techniques in Industry 4.0*,
SpringerBriefs in Applied Sciences and Technology,
https://doi.org/10.1007/978-981-13-2640-0_5

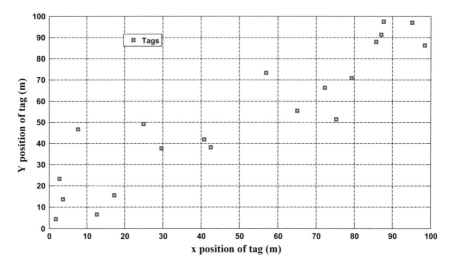

Fig. 5.1 Proposed static working area

5.2.2 Dynamic Working Area

Contrary to the static working area which tags have been pinned to their locations, in dynamic working area, tags have changing positions and have moving status.

The proposed dynamic working area in this book is the conveyor belt of a Flexible Manufacturing Systems (FMS) (see Fig. 5.2).

Fig. 5.2 FMS [1]

Using MATLAB software, the conveyor belt which has 3 m length and 2 m width has been modeled. It is assumed that the conveyor moves with constant speed of 0.3 m/s and square shape parts with length of 15 cm, which have been equipped with RFID passive tags that move along the conveyor with 15 cm distance with each other (see Fig. 5.3).

The tags are mounted on the middle of each part and moving with parts, so by defined proposed working area conditions can be interpreted in following two different scenarios:

A. One tag moves along the conveyor and its speed is equal to conveyor speed and in each 2 s, tag moves 30 cm, so to complete a tour in conveyor, tag should be detectable in 30 different positions.
B. Tags move along the conveyor and their speed is equal to conveyor speed so in each second, each tag moves 30 cm, so in whole conveyor, 30 tags move in the same time.

It should be noted that all 30 tags should be covered by RFID antennas in the same time, so it can be interpreted that 30 static tags have been deployed on the RFID network in both cases A and B.

After defining the working area and modeled it with MATLAB software the next step is defining the parameters of the proposed algorithm.

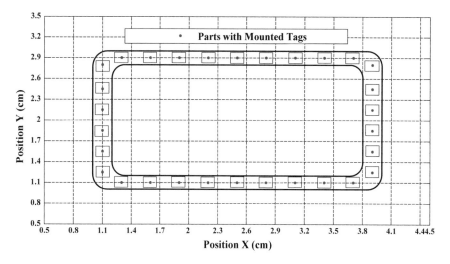

Fig. 5.3 Proposed dynamic working area

5.3   Parameters of the Proposed Hybrid Algorithm

Parameters of the proposed hybrid artificial intelligence algorithm which should be defined are divided to two sections: first, the population of the possible answers should be defined, and the next one is defining the objective function of the algorithm.

5.3.1   Population of the Possible Answers

Population of the possible answers can be defined as a matrix with m rows and n columns which each row of the matrix contains an answer of optimizing the proposed RFID network. Number of rows is a predefined arbitrary number which has been assumed in this book as 100, so the optimization process starts with 100 possible answers to RNP known as global answer space or population of the possible answers and tries to optimize it to accomplish the optimization process by searching the best answer in the global answer space.

As it has been discussed in previous chapters, optimization process starts with solving the mathematical model of RFID network and ends with choosing the best solution. Solving the mathematical model needs the number and positions of the RFID antennas which in the first iteration of the proposed optimization technique has been chosen randomly (here in this book it has been chosen to number of the tags), but in the other iterations, the required data have been provided by RPLNN algorithm. Each answer in the population of possible answers can be defined as combination of the position of deployed RFID antennas and activity status of them.

The position of antennas has been defined in Cartesian coordinates as (x, y), and activity status refers to state of the antenna which is *on* or *off*, so if an antenna activity calculated as *off*, it means that the antenna has not been deployed on the RFID network.

The positions and activity status of the antennas should be encoded in binary form as zeroes and ones at the output of RAE algorithms, which is the input of the RPLNN algorithm and should be decoded to decimal form at the input of mathematical model which is the output of the RPLNN algorithm. In this case, the number of the columns of the population matrix equals the binary string length of positions and activity status of antennas (see Fig. 5.4).

It should be noted that for the activity status of antennas in this book, 0 has been considered for non-active antennas and 1 has been taken for active antennas. An example of the possible answer encoded to binary form has been illustrated in Fig. 5.5.

$$
\begin{bmatrix}
x\ positions\ of\ antennas & y\ positions\ of\ antennas & activity\ of\ antennas \\
x\ positions\ of\ antennas & y\ positions\ of\ antennas & activity\ of\ antennas \\
x\ positions\ of\ antennas & y\ positions\ of\ antennas & activity\ of\ antennas \\
 & \cdot & \\
 & \cdot & \\
 & \cdot & \\
\end{bmatrix}
$$

Fig. 5.4 Proposed population of possible answers

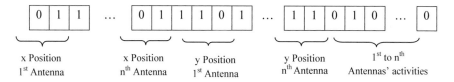

Fig. 5.5 Example of encoded answer in population of answers

5.3.2   Fitness Function

The fitness function is an evaluation tool which roles over the optimization process and can be interpreted by combination of the parameters of the RNP which should be maximized or minimized through optimization algorithm. Each row of the population of the possible answers matrix should be evaluated by the fitness function, and the possible answer which has the highest value of fitness function is the best optimization solution.

The aim of the proposed hybrid artificial intelligence algorithm in this book is establishment of an optimized RFID network by minimizing the number of deployed antennas in a manner which coverage percentage of the network reaches maximum and interference of the network reduces to minimum possible value.

The following fitness function has been proposed in this book to evaluate the performance of each possible answer of population matrix [1]:

$$
f(x_i) = \frac{100}{1 + (100 - \mathrm{COV}(i))^2} + \frac{100}{1 + \mathrm{ITF}^2(i)} + \frac{1}{1 + (\text{Number of Antennas})^2}
$$
$$
i = 1 \ldots \text{population size} \tag{5.1}
$$

The value of the proposed fitness function increases by increasing the network coverage and decreasing interference and number of deployed antennas in the network. Since coverage of the network depends on the RFID antennas, so to have covered tags at least one antenna should be deployed on the network. In this case, if the network coverage by adopting one antenna reaches 100% and interference reduces to zero, so the maximum value of the proposed fitness function value will be calculated as 201.

After defining the working area and parameters of the proposed hybrid algorithm as a part of implementation stage, the next session results of the optimizing the RNP by the proposed algorithms have been introduced in detail.

5.4 Results

The proposed hybrid artificial intelligence technique has been adopted to optimize the RNP in static and dynamic working areas and to have a better overview of the efficiency of the proposed algorithm; the performance of it has been compared with Genetic Algorithm (GA) optimization technique, which is a well-known algorithm among evolutionary optimization techniques.

The optimized positions and number of deployed RFID antennas as solution for the RNP have been calculated by both optimization paradigms for 100 iterations with the same first population of possible answers matrix, and the results have been represented and compared as follows.

5.4.1 Static Working Area

As discussed in previous sections, the proposed static working area in this book is a square room with dimension of 100×100 m^2 and has been divided into four equal segments. 20 RFID passive static tags have been randomly distributed in this working area. For the first iteration of the optimization process equal to the number of the tags, 20 RFID antennas randomly have been deployed on the network.

The predefined number of deployed antennas by implementing both optimizing algorithms has been reduced from 20 to 2 at the end of the proposed hybrid optimization process; however, this number has been calculated as 5 by GA (see Fig. 5.6). After optimizing the number of antennas, in the next step, coverage and interference of the RFID network which are the resultant of the deployed antennas have been investigated.

The results shown in Fig. 5.7 indicate that the RFID network coverage at the beginning of both optimizations processes has been calculated as 75%, and both paradigms can improve this percentage to 100%.

By investigating the network coverage results in detail, it can be noticed that however optimizing the RNP utilizing both algorithms gives solutions with full network coverage, but the proposed hybrid artificial intelligence paradigm reaches the answer in less iterations than GA.

The interference of the RFID network is resultant of collision of the deployed antennas in the network. The ITF of the proposed RFID network from -600 dBm has been reduced to -71.34 and -23.34 dBm, respectively, by adopting GA and the proposed hybrid artificial intelligence paradigms (see Fig. 5.8). Therefore, the network interference results indicate that optimizing the RFID network by the proposed

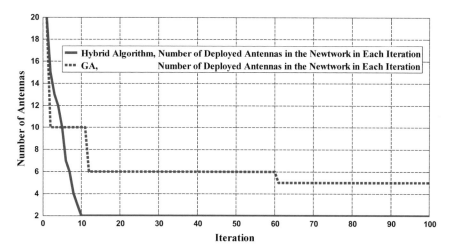

Fig. 5.6 Calculated number of deployed antennas in the network by the proposed hybrid algorithm and GA in each iteration

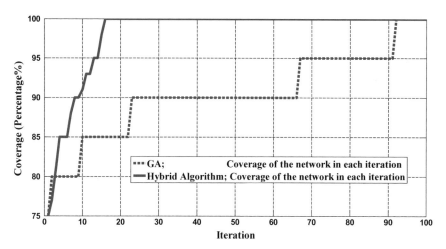

Fig. 5.7 Calculated coverage of the RFID network by the proposed hybrid algorithm and GA in each iteration

hybrid algorithm compared to GA optimization technique achieves more efficient number and positions of antennas.

Overall, the results indicate that utilizing the proposed hybrid artificial intelligence optimization technique compared to the GA optimization paradigm optimizes the RFID network and gives a solution to RNP with the fewer antenna and interference in the network.

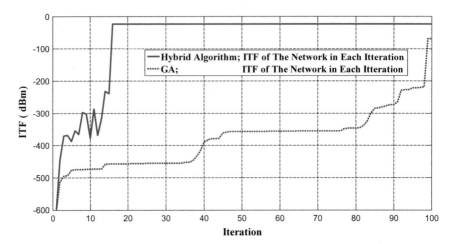

Fig. 5.8 Calculated ITF of the network by the proposed hybrid algorithm and GA in each iteration

Comparing the normalized fitness function values of two approaches which have been shown in Fig. 5.9 indicates that the proposed hybrid algorithm has superior performance than GA.

Since the RFID network can reach full network coverage in fewer iterations due to fewer deployment of antennas and less interference on the network, the proposed hybrid algorithm in compare to GA is more cost-effective and time efficient technique. The optimized solution of RNP has been shown in Fig. 5.10.

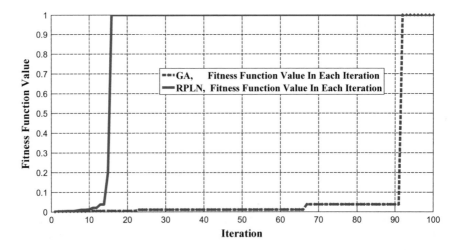

Fig. 5.9 Normalized fitness function values of the proposed hybrid algorithm and GA in each iteration

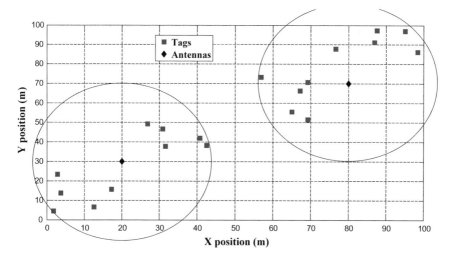

Fig. 5.10 Optimized RFID network

5.4.2   Dynamic Working Area

As discussed in previous sections, the proposed dynamic working area in this book is a conveyor belt with dimension of 3×2 m^2. 30 RFID passive dynamic tags are mounted at the center of moving parts, which have been moved along the conveyor with the constant speed of the belt that equals 0.3 m/s. For the first iteration of the optimization process equal to the number of the tags, 30 RFID antennas randomly have been deployed on the network.

The predefined number of deployed antennas by implementing both optimizing algorithms has been reduced from 30 to 1 at the end of the proposed hybrid optimization process; however, this number has been calculated as 3 by GA (see Fig. 5.11). After optimizing the number of antennas, in the next step, coverage and interference of the RFID network which are the resultant of the deployed antennas have been investigated.

The results shown in Fig. 5.12 indicate that the RFID network coverage remains as 100% from the beginning till the end of both optimization processes.

Since network coverage during both optimization processes remains as 100%, then contrary to the proposed static working area, investigating the network coverage results does not give a clue to compare the efficiency of the algorithms.

The interference of the RFID network is resultant of collision of the deployed antennas in the network. The ITF of the proposed RFID network from −1000 dBm has been reduced to 0 and −183.66 dBm, respectively, by adopting GA and the proposed hybrid artificial intelligence paradigms (see Fig. 5.13). Therefore, the network interference results indicate that optimizing the RFID network by the proposed

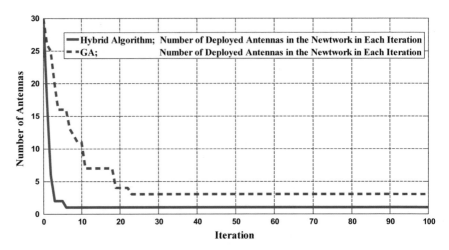

Fig. 5.11 Calculated number of deployed antennas in the network by the proposed hybrid algorithm and GA in each iteration

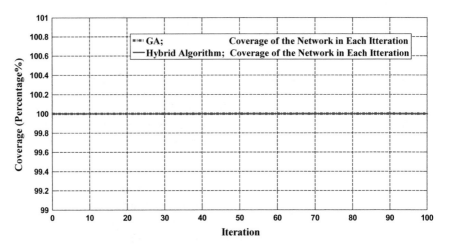

Fig. 5.12 Calculated coverage of the RFID network by the proposed hybrid algorithm and GA in each iteration

hybrid algorithm compared to GA optimization technique achieves more efficient number and positions of antennas.

Overall, the results indicate that utilizing the proposed hybrid artificial intelligence optimization technique compared to the GA optimization paradigm optimizes the RFID network and gives a solution to RNP with the fewer antenna and interference in the network.

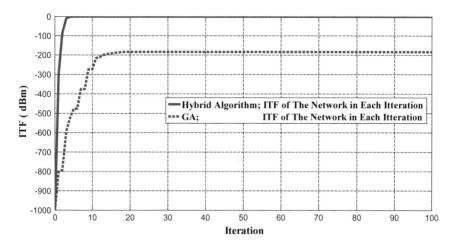

Fig. 5.13 Calculated ITF of the network by the proposed hybrid algorithm and GA in each iteration

Comparing the normalized fitness function values of two approaches which have been shown in Fig. 5.14 indicates that the proposed hybrid algorithm has superior performance than GA.

Since the RFID network can reach full network coverage in fewer iterations due to fewer deployment of antennas and less interference on the network, the proposed hybrid algorithm in compare to GA is more cost-effective and time efficient technique. The optimized solution of RNP has been shown in Fig. 5.15.

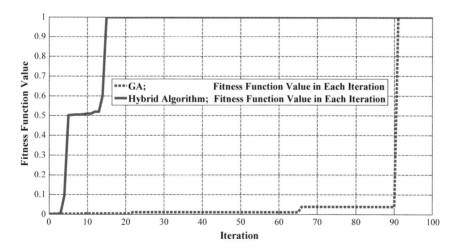

Fig. 5.14 Normalized fitness function values of the proposed hybrid algorithm and GA in each iteration

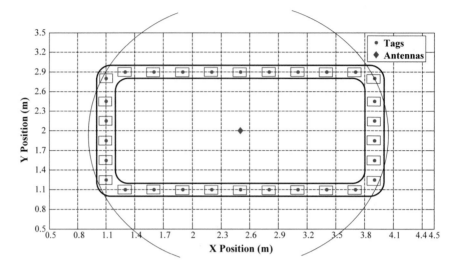

Fig. 5.15 Optimized RFID network

5.5 Conclusion

In this book, by combining optimization paradigms which have been known as RPLNN and RAE, a novel efficient hybrid artificial intelligence optimization technique is introduced to deal with a complex RFID network planning problem. This proposed optimization technique can adjust any number of embedded RFID antennas in the network. The simulation-based performance assessment has been performed for investigating the effectiveness of the hybrid algorithm over the GA. The simulation results show that the proposed hybrid algorithm has superior performance over the GA. It has been observed that the hybrid optimization technique for optimizing the RNP is more enhanced than GA as:

1. The proposed hybrid algorithm has faster convergence speed and lesser iterations than GA.
2. The proposed hybrid algorithm has less complexity in computations than GA.
3. The results given by the proposed hybrid algorithm are more precise and cost-effective.

In future studies, with an innovation in the proposed mathematical models of the RNP, it will be possible to model an RFID network that involves more criteria such as the quality of the network. A valuable future study would involve the utilization of more optimization techniques for static and dynamic networks to generate RFID networks.

Reference

1. A. Azizi, Introducing a novel hybrid artificial intelligence algorithm to optimize network of industrial applications in modern manufacturing. Complexity **2017** (2017)

Printed in the United States
By Bookmasters